div, grad, curl, and all that

third edition

div, grad,

an informal text

h. m. schey

curl,
and all that

on vector calculus

third edition

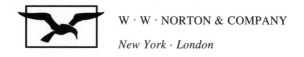

W · W · NORTON & COMPANY

New York · London

The text of this book is composed in Times Roman with the display set in Optima. Composition by University Graphics, Inc.

Library of Congress Cataloging-in-Publication Data
Schey, H. M. (Harry Moritz), 1930–
    Div, grad, curl, and all that: an informal text on vector
calculus / H. M. Schey.—3rd ed.
        p.   cm.
    Includes bibliographical references (pp. 160–61 and
index.
    1. Vector analysis. I. Title.
    QA433.S28   1996
    515′.63—dc20   96-4942

ISBN 0-393-96997-5 (pbk.)

W. W. Norton & Company, Inc., 500 Fifth Avenue, New York, N.Y. 10110
www.wwnorton.com

W. W. Norton & Company Ltd., Castle House, 75/76 Wells Street, London W1T 3QT

Zusammengestohlen aus Verschiedenem diesem und jenem.

Ludwig van Beethoven

Contents

Contents

Preface to the Third Edition

If it ain't broke, don't fix it.

Anonymous

This new edition constitutes a fine-tuning of its predecessor. Several new problems have been added, two other problems awkwardly worded in the earlier editions have been revised, and a diagram has been corrected. The major change involves replacing the operators div, grad, and curl by the appropriate expressions using the ∇ operator, to bring the text into conformity with modern notational practice. I have, however, resisted retitling the book $\nabla\cdot$, ∇, $\nabla \times$, and All That.

I wish to express my gratitude to Richard Liu, Stephen Nettel, and Sally Seidel for their useful reviews of the previous edition. I take particular pleasure in thanking those of my readers who over the years have been good enough to send me comments, criticisms, and suggestions which have contributed significantly to the quality of the text.

Chapter I

Introduction, Vector Functions, and Electrostatics

One lesson, Nature, let me learn of thee.

Matthew Arnold

Introduction

In this text the subject of vector calculus is presented in the context of simple electrostatics. We follow this procedure for two reasons. First, much of vector calculus was invented for use in electromagnetic theory and is ideally suited to it. This presentation will therefore show what vector calculus is and at the same time give you an idea of what it's for. Second, we have a deep-seated conviction that mathematics—in any case *some* mathematics—is best discussed in a context which is not exclusively mathematical. Thus, we will soft pedal mathematical rigor, which

1

we think is an obstacle to learning this subject on a first exposure
to it, and appeal as much as possible to physical and geometric
intuition.

Now, if you want to learn vector calculus but know little or
nothing about electrostatics, you needn't be put off by our
approach; no very great knowledge of physics is required to read
and understand this text. Only the simplest features of electrostatics are involved, and these are presented in a few pages near
the beginning. It should not be an impediment to anyone. In fact,
the entire discussion is based on a search for a convenient method
of finding the electrostatic field given the distribution of electric
charges which produce it. This is the thread which runs through,
and unifies, our presentation, so that as a bare minimum all you
really need do is take our word for the fact that the electric field
is an important enough quantity to warrant spending some time
and effort in setting up a general method for calculating it. In the
process, we hope you will learn the elements of vector calculus.

Having said what you do *not* need to know, we must now say
what you *do* need to know. To begin with, you should, of course,
be fluent in elementary calculus. Beyond that you should know
how to work with functions of several variables, partial derivatives, and multiple (double and triple) integrals.[1] Finally, you
must know something about vectors. This, however, is a subject
of which too many writers and teachers have made heavy
weather. What you should know about it can be listed quickly:
definition of vector, addition and subtraction of vectors, multi-
plication of vectors by scalars, dot and cross products, and finally,
resolution of vectors into components. An hour's time with any
reasonable text on the subject should provide you with all you
need to know of it to follow this text.

Vector Functions

One of the most important quantitites we deal with in the study
of electricity is the electric field, and much of our presentation
will make use of this quantity. Since the electric field is an example of what we call a vector function, we begin our discussion
with a brief résumé of the function concept.

A function of one variable, generally written $y = f(x)$, is a *rule*

[1] Differential equations are used in one section of this text. The section is not
essential and may be omitted if the mathematics is too frightening.

Vector Functions which tells us how to associate two numbers x and y; given x, the function tells us how to determine the associated value of y. Thus, for example, if $y = f(x) = x^2 - 2$, then we calculate y by squaring x and then subtracting 2. So, if $x = 3$,

$$y = 3^2 - 2 = 7.$$

Functions of more than one variable are also rules for associating sets of numbers. For example, a function of three variables designated $w = F(x, y, z)$ tells how to assign a value to w given x, y, and z. It is helpful to view this concept geometrically; taking (x, y, z) to be the Cartesian coordinates of a point in space, the function $F(x, y, z)$ tells us how to associate a number with each point. As an illustration, a function $T(x, y, z)$ might give the temperature at any point (x, y, z) in a room.

The functions so far discussed are *scalar* functions; the result of "plugging" x in $f(x)$ is the scalar $y = f(x)$. The result of "plugging" the three numbers x, y, and z in $T(x, y, z)$ is the temperature, a scalar. The generalization to vector functions is straightforward. A vector function (in three dimensions) is a rule which tells us how to associate a *vector* with each point (x, y, z). An example is the velocity of a fluid. Designating this function $\mathbf{v}(x, y, z)$, it specifies the *speed* of the fluid as well as the *direction* of flow at the point (x, y, z). In general, a vector function $\mathbf{F}(x, y, z)$ specifies a *magnitude* and a *direction* at every point (x, y, z) in some region of space. We can picture a vector function as a collection of arrows (Figure I–1), one for each point (x, y, z).

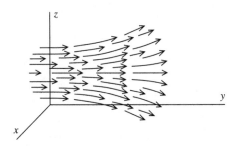

Figure I–1

The direction of the arrow at any point is the direction specified by the vector function, and its length is proportional to the magnitude of the function.

A vector function, like any vector, can be resolved into com-

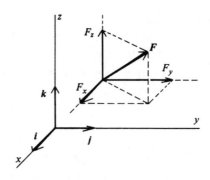

Figure I–2

ponents, as in Figure I–2. Letting **i**, **j**, and **k** be unit vectors along the x-, y-, and z-axes, respectively, we write

$$\mathbf{F}(x, y, z) = \mathbf{i}F_x(x, y, z) + \mathbf{j}F_y(x, y, z) + \mathbf{k}F_z(x, y, z).$$

The three quantities F_x, F_y, and F_z, which are themselves scalar functions of x, y, and z, are the three Cartesian components of the vector function $\mathbf{F}(x, y, z)$ in some coordinate system.[2]

An example of a vector function (in two dimensions for simplicity) is provided by

$$\mathbf{F}(x, y) = \mathbf{i}x + \mathbf{j}y,$$

which is illustrated in Figure I–3. You probably recognize this

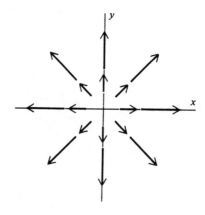

Figure I–3

[2] Some writers use subscripts to indicate the partial derivative; for example, $F_x = \partial F/\partial x$. We shall *not* adopt such notation here; our subscripts will always denote the vector component.

function as the position vector **r**. Each arrow in the figure is in the *radial* direction (that is, directed along a line emanating from the origin) and has a length equal to its distance from the origin.[3] A second example,

$$G(x, y) = \frac{-\mathbf{i}y + \mathbf{j}x}{\sqrt{x^2 + y^2}},$$

is shown in Figure I–4. You should verify for yourself that for this vector function all the arrows are in the tangential direction

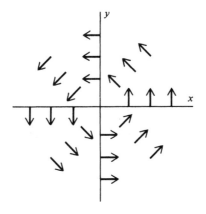

Figure I–4

(that is, each is tangent to a circle centered at the origin) and all have the same length.

Electrostatics

We shall base our discussion of electrostatics on three experimental facts. The first of these facts is the existence of electric charge itself. There are two kinds of charge, positive and negative, and every material body contains electric charge,[4] although

---

[3] Note that by convention an arrow is drawn with its tail, *not its head,* at the point where the vector function is evaluated.
[4] Purists will point out that neutrons, neutral pi mesons, neutrinos, and the like do not contain charge.

Figure I–5

often the positive and negative charges are present in equal amounts so that there is zero *net* charge.

The second fact is called Coulomb's law, after the French physicist who discovered it. This law states that the electrostatic force between two charged particles (a) is proportional to the product of their charges, (b) is inversely proportional to the square of the distance between them, and (c) acts along the line joining them. Thus, if q_0 and q are the charges of two particles a distance r apart (Figure I–5), then the force on q_0 due to q is

$$\mathbf{F} = K \frac{qq_0}{r^2} \hat{\mathbf{u}},$$

where $\hat{\mathbf{u}}$ is a unit vector (that is, a vector a length 1) pointing from q to q_0, and K is a constant of proportionality. In this text we'll use rationalized MKS units. In that system length, mass, and time are measured in meters, kilograms, and seconds, respectively, and electric charge in coulombs. With this choice of units $K = (1/4\pi\epsilon_0)$, where the constant ϵ_0, called the permittivity of free space, has the value 8.854×10^{-12} coulombs^2 per newton-meters^2, and Coulomb's law reads

$$\mathbf{F} = \frac{1}{4\pi\epsilon_0} \frac{qq_0}{r^2} \hat{\mathbf{u}}. \tag{I-1}$$

You should convince yourself that the familiar rule "like charges repel, unlike charges attract" is built into this formula.

The third and last fact is called the principle of superposition. If $\mathbf{F}_1$ is the force exerted on q_0 by q_1 when there are no other charges nearby, and $\mathbf{F}_2$ is the force exerted on q_0 by q_2 when there are no other charges nearby, then the principle of superposition says that the net force exerted on q_0 by q_1 and q_2 when they are both present is the vector sum $\mathbf{F}_1 + \mathbf{F}_2$. This is a deeper statement than it appears at first glance. It says not merely that electrostatic forces add vectorially (*all* forces add vectorially), but that the force between two charged particles is not modified by the presence of other charged particles.

We now introduce a vector function of position which will play a leading role in our discussion. It is the electrostatic field, denoted $\mathbf{E}(\mathbf{r})$ and defined by the equation $\mathbf{E}(\mathbf{r}) = \mathbf{F}(\mathbf{r})/q_0$, or $\mathbf{F}(\mathbf{r}) = q_0\mathbf{E}(\mathbf{r})$. That is, the electrostatic field is the force per unit charge. From Equation (I–1) we have

$$\mathbf{E}(\mathbf{r}) = \frac{\mathbf{F}(\mathbf{r})}{q_0} = \frac{1}{4\pi\epsilon_0}\frac{q}{r^2}\,\hat{\mathbf{u}}. \qquad (\text{I–2})$$

This is the electrostatic field at $\mathbf{r}$ due to the charge q.

A natural extension of these ideas is the following. Suppose we have a group of N charges with q_1 situated at $\mathbf{r}_1$, q_2 at $\mathbf{r}_2, \ldots,$ q_N at $\mathbf{r}_N$. Then the electrostatic force these charges exert on a charge q_0 situated at r is

$$\mathbf{F}(\mathbf{r}) = \frac{1}{4\pi\epsilon_0}\sum_{l=1}^{N}\frac{q_0q_l}{|\mathbf{r} - \mathbf{r}_l|^2}\,\hat{\mathbf{u}}_l, \qquad (\text{I–3})$$

where $\hat{\mathbf{u}}_l$ is the unit vector pointing from $\mathbf{r}_l$ to $\mathbf{r}$. From Equation (I–3) we have

$$\mathbf{E}(\mathbf{r}) = \frac{1}{4\pi\epsilon_0}\sum_{l=1}^{N}\frac{q_l}{|\mathbf{r} - \mathbf{r}_l|^2}\,\hat{\mathbf{u}}_l. \qquad (\text{I–4})$$

This is the electrostatic field at $\mathbf{r} = \mathbf{i}x + \mathbf{j}y + \mathbf{k}z$ produced by the charges q_l at $\mathbf{r}_l$ ($l = 1, 2, \ldots, N$). Equation (I–4) says that the field due to a group of charges is the vector sum of the fields each produces alone. That is, superposition holds for fields as well as forces. You may think of the region of space in the vicinity of a charge or group of charges as "pervaded" by an electrostatic field; the net electrostatic force exerted by those charges on a charge q at a point $\mathbf{r}$ is then $q\mathbf{E}(\mathbf{r})$.

You may be a bit mystified about our bothering to introduce a new vector function, the electrostatic field, which differs in an apparently trivial way from the electrostatic force. There are two major reasons for doing this. First, in electrostatics we are interested in the effect that a given set of charges produces on another set. This problem can be conveniently divided into two parts by introducing the electrostatic field, for then we can (a) calculate the field due to a given distribution of charges without worrying about the effect these charges have on *other* charges in the vicinity and (b) calculate the effect a given field has on charges placed in it without worrying about the distribution of charges that produced the field. In this book we will be concerned with the first of these.

The second reason for introducing the electrostatic field is more basic. It turns out that all classical electromagnetic theory can be codified in terms of four equations, called Maxwell's equations, which relate fields (electric and magnetic) to each other and to the charges and currents which produce them. Thus, electromagnetism is a *field theory* and the electric field ultimately plays a role and assumes an importance which far transcends its simple elementary definition as "force per unit charge."

Very often it is convenient to treat a distribution of electric charge as if it were continuous. To do this, we proceed as follows. Suppose in some region of space of volume ΔV the total electric charge is ΔQ. We define the *average charge density* in ΔV as

$$\bar{\rho}_{\Delta V} \equiv \frac{\Delta Q}{\Delta V}. \qquad (I\text{--}5)$$

Using this, we can define the charge density at the point (x, y, z), denoted $\rho(x, y, z)$, by taking the limit of $\bar{\rho}_{\Delta V}$ as ΔV shrinks down about the point (x, y, z):

$$\rho(x, y, z) \equiv \lim_{\substack{\Delta V \to 0 \\ \text{about } (x,y,z)}} \frac{\Delta Q}{\Delta V} = \lim_{\substack{\Delta V \to 0 \\ \text{about } (x,y,z)}} \bar{\rho}_{\Delta V}. \qquad (I\text{--}6)$$

The electric charge in some region of volume V can then be expressed as the triple integral of $\rho(x, y, z)$ over the volume V; that is,

$$Q = \iiint_V \rho(x, y, z) \, dV.$$

In much the same way it follows that for a continuous distribution of charges, Equation (I–4) is replaced by

$$\mathbf{E}(\mathbf{r}) = \frac{1}{4\pi\epsilon_0} \iiint_V \frac{\rho(\mathbf{r}')\hat{\mathbf{u}}(\mathbf{r}')}{|\mathbf{r} - \mathbf{r}'|^2} \, dV'. \qquad\qquad \text{(I–7)}$$

PROBLEMS

I–1 Using arrows of the proper magnitude and direction, sketch each of the following vector functions:

(a) $\mathbf{i}y + \mathbf{j}x$. (e) $\mathbf{j}x$.
(b) $(\mathbf{i} + \mathbf{j})/\sqrt{2}$. (f) $(\mathbf{i}y + \mathbf{j}x)/\sqrt{x^2 + y^2}$, $(x, y) \neq (0, 0)$.
(c) $\mathbf{i}x - \mathbf{j}y$. (g) $\mathbf{i}y + \mathbf{j}xy$.
(d) $\mathbf{i}y$. (h) $\mathbf{i} + \mathbf{j}y$.

I–2 Using arrows, sketch the electric field of a unit positive charge situated at the origin. [*Note:* You may simplify the problem by confining your sketch to one of the coordinate planes. Does it matter which plane you choose?]

I–3 (a) Write a formula for a vector function in two dimensions which is in the positive radial direction and whose magnitude is 1.

(b) Write a formula for a vector function in two dimensions whose direction makes an angle of 45° with the x-axis and whose magnitude at any point (x, y) is $(x + y)^2$.

(c) Write a formula for a vector function in two dimensions whose direction is tangential (in the sense of the example on page 5) and whose magnitude at any point (x, y) is equal to its distance from the origin.

(d) Write a formula for a vector function in three dimensions which is in the positive radial direction and whose magnitude is 1.

I–4 An object moves in the xy-plane in such a way that its position vector $\mathbf{r}$ is given as a function of time t by

$$\mathbf{r} = \mathbf{i}a \cos \omega t + \mathbf{j}b \sin \omega t,$$

where a, b, and ω are constants.

(a) How far is the object from the origin at any time t?
(b) Find the object's velocity and acceleration as functions of time.
(c) Show that the object moves on the elliptical path

$$\left(\frac{x}{a}\right)^2 + \left(\frac{y}{b}\right)^2 = 1.$$

I–5 A charge $+ 1$ is situated at the point $(1, 0, 0)$ and a charge -1 is situated at the point $(-1, 0, 0)$. Find the electric field of these two charges at an arbitrary point $(0, y, 0)$ on the y-axis.

I–6 Instead of using arrows to represent vector functions (as in Problems I–1 and I–2), we sometimes use families of curves called *field lines.* A curve $y = y(x)$ is a field line of the vector function $\mathbf{F}(x, y)$ if at each point (x_0, y_0) on the curve, $\mathbf{F}(x_0, y_0)$ is tangent to the curve (see the figure).

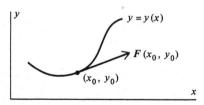

(a) Show that the field lines $y = y(x)$ of a vector function

$$\mathbf{F}(x, y) = \mathbf{i}F_x(x, y) + \mathbf{j}F_y(x, y)$$

are solutions of the differential equation

$$\frac{dy}{dx} = \frac{F_y(x, y)}{F_x(x, y)}.$$

(b) Determine the field lines of each of the functions of Problem I–1. Draw the field lines and compare with the arrow diagrams of Problem I–1.

Chapter II

Surface Integrals and the Divergence

Oh, could I flow like thee, and make
thy stream
My great example . . .

Sir John Denham

Gauss' Law

Since the electrostatic field is so important a quantity in electrostatics, it follows that we need some convenient way to find it, given a set of charges. At first glance it might appear that we solved this problem before we even stated it, for, after all, do not Equations (I–4) and (I–7) provide us with a means of finding **E**? The answer is, in general, no. Unless there are very few charges in the system and/or they are arranged simply or very symmetrically, the sum in Equation (I–4) and the integral in Equation (I–7) are usually prohibitively difficult—and frequently impossible—to perform. Thus, these two equations provide what is

11

usually only a "formal" solution[1] to the problem, not a practical one, and we must cast about for some other way to calculate the field **E**.

In the course of this casting about, we come inevitably to that remarkable relation known as Gauss' law. We say "inevitably" because it is hard to think of any other expression in elementary electricity and magnetism containing the electric field [apart, of course, from Equations (I–4) and (I–7), which we have already rejected]. Gauss' law is

$$\iint_S \mathbf{E} \cdot \hat{\mathbf{n}} \, dS = \frac{q}{\epsilon_0} . \qquad \text{(II–1)}$$

If you don't understand this equation, don't panic. The left-hand side of this equation is an example of what is called a surface integral, an important concept in vector calculus and one that is probably new to you. The integrand of this integral is the dot product of the electric field and the quantity $\hat{\mathbf{n}}$, which is called a "unit normal vector" and is probably also unfamiliar. We are about to discuss both surface integrals and unit normal vectors in excruciating detail, and one of our main reasons for quoting Gauss' law at this point in our narrative is to motivate this discussion.

We won't stop here to derive Gauss' law, since the derivation wouldn't mean much to you until you have read the next few sections. Then you can consult almost any text on electricity and magnetism for the gory details. And if you can contain yourself, wait until we've discussed the divergence theorem (pages 44–52), after which you will be able to derive Gauss' law easily (see Problem II–27).

The Unit Normal Vector

One of the factors in the integrand in Gauss' law [Equation (II–1)] is a quantity designated $\hat{\mathbf{n}}$ and called the unit normal vector. This quantity is part of the integrand in most if not all of the surface integrals we'll encounter; furthermore, as we'll see, it plays an important role in the evaluation of surface integrals even

12

---

[1] The word "formal" in this context is a euphemism for "useless."

when it does not appear explicitly. Thus, before discussing surface integrals themselves, we'll dispose of the questions of how this vector function is defined and calculated.

The word "normal" in the present context means, loosely speaking, "perpendicular." Thus, a vector **N** normal to the *xy*-plane is clearly one parallel to the *z*-axis (Figure II–1), while a

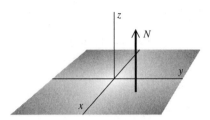

Figure II–1

vector normal to a spherical surface must be in the radial direction (Figure II–2). To give a precise definition of a vector normal to

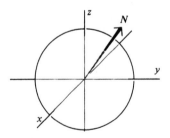

Figure II–2

a surface, consider an arbitrary surface *S* as shown in Figure II–3. Construct two noncollinear vectors **u** and **v** tangent to *S* at some point *P*. A vector **N** which is perpendicular to both **u** and **v** at *P* is, by definition, normal to *S* at *P*. Now, as we know, the

13

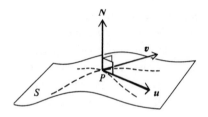

Figure II–3

vector product of **u** and **v** has precisely this property; it is perpendicular to both **u** and **v**. Thus, we may write $N = u \times v$. To
make this a *unit* vector (that is, one whose length is 1) is simple:
we just divide **N** by its magnitude N. Thus,

$$\hat{n} = \frac{N}{N} = \frac{u \times v}{|u \times v|}$$

is a unit vector normal to S at P.

To find an expression for $\hat{n}$, we consider some surface S given
by the equation $z = f(x, y)$; see Figure II–4. Following the pro-

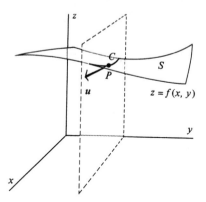

Figure II–4

cedure suggested by the above discussion, we'll find two vectors
u and **v** whose cross product will yield the required normal vector
$\hat{n}$. For this purpose let's construct a plane through a point P on
S and parallel to the xz-plane, as shown in Figure II–4. This plane
intersects the surface S in a curve C. We construct the vector **u**
tangent to C at P and having an x-component of arbitrary length

u_x. The z-component of $\mathbf{u}$ is $(\partial f/\partial x)u_x$; in this expression we use the fact that the slope of $\mathbf{u}$ is, by construction, the same as that

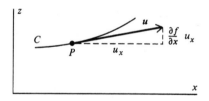

Figure II–5

of the surface S in the x-direction (see Figure II–5). Thus,

$$\mathbf{u} = \mathbf{i}u_x + \mathbf{k}\left(\frac{\partial f}{\partial x}\right)u_x = \left[\mathbf{i} + \mathbf{k}\left(\frac{\partial f}{\partial x}\right)\right]u_x. \qquad \text{(II–2)}$$

To find $\mathbf{v}$, the second of our two vectors, we pass another plane through the point P on S, but in this case parallel to the yz-plane (Figure II–6). It intersects S in a curve C', and the vector $\mathbf{v}$ can

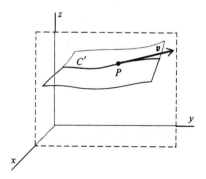

Figure II–6

now be constructed tangent to C' at P with a y-component of arbitrary length v_y. Arguing as above, we have

$$\mathbf{v} = \mathbf{j}v_y + \mathbf{k}\left(\frac{\partial f}{\partial y}\right)v_y = \left[\mathbf{j} + \mathbf{k}\left(\frac{\partial f}{\partial y}\right)\right]v_y. \qquad \text{(II–3)}$$

Using the two vectors $\mathbf{u}$ and $\mathbf{v}$ as given in Equations (II–2) and (II–3), we now construct their cross product. The result,

15

$$\mathbf{u} \times \mathbf{v} = \left[-\mathbf{i} \left(\frac{\partial f}{\partial x} \right) - \mathbf{j} \left(\frac{\partial f}{\partial y} \right) + \mathbf{k} \right] u_x v_y,$$

is a vector, which as we stated above, is normal to S at P. To make a *unit* vector of this, we divide it by its magnitude to get

$$\hat{\mathbf{n}}(x, y, z) = \frac{\mathbf{u} \times \mathbf{v}}{|\mathbf{u} \times \mathbf{v}|} = \frac{-\mathbf{i} \dfrac{\partial f}{\partial x} - \mathbf{j} \dfrac{\partial f}{\partial y} + \mathbf{k}}{\sqrt{1 + \left(\dfrac{\partial f}{\partial x} \right)^2 + \left(\dfrac{\partial f}{\partial y} \right)^2}}. \qquad \text{(II–4)}$$

This, then, is the unit vector normal to the surface $z = f(x, y)$ at the point (x, y, z) on the surface.[2] Note that it is independent of the two arbitrary quantities u_x and v_y.

A couple of examples may be in order here. First a trivial one: What is the unit vector normal to the xy-plane? The answer, of course, is $\mathbf{k}$ (see Figure II–1). Let's see how Equation (II–4) provides us with this answer. The equation of the xy-plane is

$$z = f(x, y) = 0,$$

whence we have the profound observations

$$\partial f / \partial x = 0 \qquad \text{and} \qquad \partial f / \partial y = 0.$$

Substituting these in Equation (II–4), we get $\hat{\mathbf{n}} = \mathbf{k}/\sqrt{1} = \mathbf{k}$, as advertised.

As a second example, consider the sphere of radius 1 centered at the origin (Figure II–2). Its upper hemisphere is given by

$$z = f(x, y) = (1 - x^2 - y^2)^{1/2},$$

whence

---

[2] The uniqueness of our result [Equation (II–4)] may be questioned on two counts. The first of these is a sign ambiguity: If $\hat{\mathbf{n}}$ is a unit normal vector, so is $-\hat{\mathbf{n}}$. The matter of which sign to use is discussed below. The second question arises from the fact that the two tangent vectors $\mathbf{u}$ and $\mathbf{v}$ used in determining $\hat{\mathbf{n}}$ are rather special, since each is parallel to one of the coordinate planes. Would we get the same result using two arbitrary tangent vectors? This issue is considered in Problem IV–26, where it is shown that $\hat{\mathbf{n}}$ as given by Equation (II–4) is, apart from sign, indeed unique.

**Definition of
Surface Integrals**

$$\frac{\partial f}{\partial x} = -\frac{x}{z} \quad \text{and} \quad \frac{\partial f}{\partial y} = -\frac{y}{z}.$$

Using these in Equation (II–4) leads to

$$\hat{\mathbf{n}} = \frac{\dfrac{\mathbf{i}x}{z} + \dfrac{\mathbf{j}y}{z} + \mathbf{k}}{\sqrt{\dfrac{x^2}{z^2} + \dfrac{y^2}{z^2} + 1}} = \frac{\mathbf{i}x + \mathbf{j}y + \mathbf{k}z}{\sqrt{x^2 + y^2 + z^2}} = \mathbf{i}x + \mathbf{j}y + \mathbf{k}z,$$

where we have used the equation of the unit sphere $x^2 + y^2 + z^2 = 1$. This is, as expected, a vector in the radial direction (see Figure II–2). To show that its length is 1, we observe that $\hat{\mathbf{n}} \cdot \hat{\mathbf{n}} = x^2 + y^2 + z^2 = 1$.

With the matter of the unit normal vector now disposed of, we turn to our next task, a discussion of surface integrals.

Definition of Surface Integrals

We now define the surface integral of the normal component of a vector function $\mathbf{F}(x, y, z)$. This quantity is denoted by

$$\iint_S \mathbf{F} \cdot \hat{\mathbf{n}} \, dS, \tag{II–5}$$

and as you can see, Gauss' law [Equation (II–1)] is expressed in terms of just such an integral. Let $z = f(x, y)$ be the equation of some surface. We'll consider a limited portion of this surface, which we designate S (see Figure II–7). Our first step in formu-

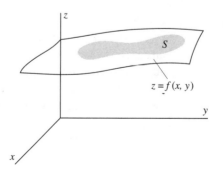

Figure II–7

lating the definition of the surface integral (II–5) is to approximate S by a polyhedron consisting of N plane faces each of which is tangent to S at some point. Figure II–8 shows how this approx-

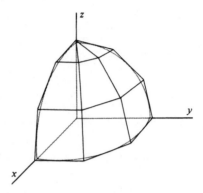

Figure II–8

imating polyhedron might look for an octant of a spherical shell. We concentrate our attention on one of these plane faces, say the lth one (Figure II–9). Let its area be denoted ΔS_l and let

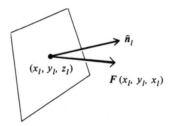

Figure II–9

(x_l, y_l, z_l) be the coordinates of the point at which the face is tangent to the surface S. We evaluate the function $\mathbf{F}$ at this point and then form its dot product with $\hat{\mathbf{n}}_l$, the unit vector normal to the lth face. The resulting quantity, $\mathbf{F}(x_l, y_l, z_l) \cdot \hat{\mathbf{n}}_l$, is then multiplied by the area ΔS_l of the face to give

$$\mathbf{F}(x_l, y_l, z_l) \cdot \hat{\mathbf{n}}_l \, \Delta S_l.$$

We carry out this same process for each of the N faces of the approximating polyhedron and then form the sum over all N faces:

$$\sum_{l=1}^{N} \mathbf{F}(x_l, y_l, z_l) \cdot \hat{\mathbf{n}}_l \, \Delta S_l.$$

The surface integral (II–5) is defined as the *limit* of this sum as the number of faces, N, approaches infinity and the area of *each* face approaches zero.[3] Thus,

$$\iint_S \mathbf{F} \cdot \hat{\mathbf{n}} \, dS = \lim_{\substack{N \to \infty \\ \text{each } \Delta S_l \to 0}} \sum_{l=1}^{N} \mathbf{F}(x_l, y_l, z_l) \cdot \hat{\mathbf{n}}_l \, \Delta S_l. \quad \text{(II–6)}$$

If we want to cross all the *t*'s and dot all the *i*'s, this integral, strictly speaking, should be written

$$\iint_S \mathbf{F}(x, y, z) \cdot \hat{\mathbf{n}}(x, y, z) \, dS$$

since both $\mathbf{F}$ and $\hat{\mathbf{n}}$ are in general functions of position. We prefer, and where possible will use, the less cluttered notation

$$\iint_S \mathbf{F} \cdot \hat{\mathbf{n}} \, dS$$

with the arguments of the functions understood.

The surface S over which we integrate a surface integral can be one of two kinds: closed or open. A closed surface, such as a spherical shell, divides space into two parts, an inside and an outside, and to get from inside to outside, you must go *through* the surface. An open surface, such as a flat piece of paper, does not have this property; it is possible to get from one side of the sheet to the other without going through it. The definition of surface integrals given in Equation (II–6) applies equally well to both closed and open surfaces. However, the surface integral is not well-defined until we specify which of the two possible directions of the normal we are to use (see Figure II–10). In the case of an open surface, the direction must be given as part of the statement of the problem. In the case of a closed surface, there is a gentlemen's agreement which specifies the direction once-

---

[3] The statement ''each $\Delta S_l \to 0$'' is not quite precise. The area of a rectangular patch, for example, might tend to zero because its width decreases while its length remains fixed. This would not be acceptable. Here and elsewhere we must interpret ''each $\Delta S_l \to 0$'' to mean that *all* linear dimensions of the *l*th patch tend to zero.

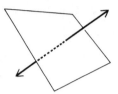

Figure II–10

and-for-all: the normal is chosen so that it points *outward* from the volume enclosed by the surface.

The integral in Gauss' law [Equation (II–1)] is taken over a closed surface. Gauss' law, in fact, says that the surface integral of the normal component of the electric field over a closed surface is equal to the total (net) charge enclosed by the surface, divided by ϵ_0. Below (pages 32–36 and Problems II–11, II–12, and II–13) we'll see how, when the charges are arranged neatly and symmetrically, Gauss' law can be used to determine the electric field. But the thrust of our whole discussion will be to subject Gauss' law to a series of harrowing adventures which eventually transform it into an expression useful for finding **E** even when we don't have symmetry to help us.

Sometimes we encounter surface integrals which are a little simpler than the kind we've just defined, although basically they are almost the same. These are surface integrals of the form

$$\iint_S G(x, y, z) \, dS, \tag{II–7}$$

where the integrand $G(x, y, z)$ is a *given* scalar function rather than the dot product of two vector functions as in (II–5) and (II–6). We go about defining this kind of surface integral much as we did above: we approximate S by a polyhedron, form the product $G(x_l, y_l, z_l) \, \Delta S_l$, sum over all faces, and then take the limit:

$$\iint_S G(x, y, z) \, dS = \lim_{\substack{N \to \infty \\ \text{each } \Delta S_l \to 0}} \sum_{l=1}^{N} G(x_l, y_l, z_l) \, \Delta S_l. \tag{II–8}$$

As an example of this kind of surface integral, suppose we have a surface of negligible thickness with surface density (that is, mass per unit area) $\sigma(x, y, z)$, and we wish to determine its total mass. Approximating this surface by a polyhedron as above, we recognize that $\sigma(x_l, y_l, z_l) \, \Delta S_l$ is approximately the mass of the *l*th face of the polyhedron and that

$$\sum_{l=1}^{N} \sigma(x_l, y_l, z_l) \, \Delta S_l$$

is approximately the mass of the entire surface. Taking the limit

$$\lim_{\substack{N \to \infty \\ \text{each } \Delta S_l \to 0}} \sum_{l=1}^{N} \sigma(x_l, y_l, z_l) \, \Delta S_l = \iint_S \sigma(x, y, z) \, dS,$$

we get the total mass of the surface.

An example of an even simpler surface integral of this kind is

$$\iint_S dS.$$

This integral is taken as the definition of the surface area of *S*.

Evaluating Surface Integrals

Now that we have defined surface integrals, we must develop methods to evaluate them, and that will be our task here. For simplicity we'll deal with surface integrals of the form (II–7), where the integrand is a given scalar function, rather than the slightly more complicated form (II–5). There will be no loss of generality in doing this for all our results can be made to apply to integrals of the form (II–5) just by replacing $G(x, y, z)$ everywhere by $\mathbf{F}(x, y, z) \cdot \hat{\mathbf{n}}$.

To evaluate the integral

$$\iint_S G(x, y, z) \, dS$$

over a portion *S* of the surface $z = f(x, y)$ (see Figure II–11), we go back to the definition of the surface integral [Equation (II–8)]. Our strategy will be to relate ΔS_l to the area ΔR_l of its projection on the *xy*-plane, as shown in Figure II–12. Doing so, as we'll see, will enable us to express the surface integral over *S* in terms of an ordinary double integral over *R*, which is the projection of *S* on the *xy*-plane, as shown in Figure II–11.

21

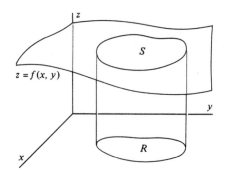

Figure II–11

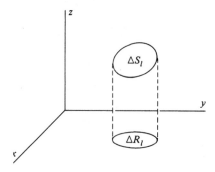

Figure II–12

Relating ΔS_l to ΔR_l is not difficult if we recall that ΔS_l (like the area of any plane surface) can be approximated to any desired degree of accuracy by a set of rectangles as shown in Figure II–13. For this reason we need only find the relation between the

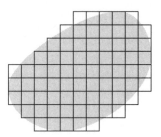

Figure II–13

22

area of a rectangle and its projection on the xy-plane. Thus, con-

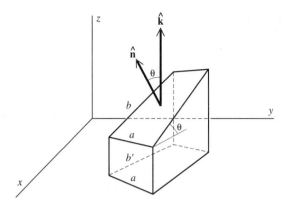

Figure II–14

sider a rectangle so oriented that one pair of its sides is parallel to the xy-plane (Figure II–14). If we call the lengths of these sides a, it's clear that their projections on the xy-plane also have length a. But the other pair of sides, of length b, have projections of length b', and in general, b and b' are *not* equal. Thus, to relate the area of the rectangle ab to the area of its projection ab', we must express b in terms of b'. This is easy to do, for if θ is the angle shown in Figure II–14, we have $b = b'/\cos\theta$, and so

$$ab = \frac{ab'}{\cos\theta}.$$

If we let $\hat{\mathbf{n}}$ denote the unit vector normal to our rectangle, then we can readily convince ourselves that $\cos\theta = \hat{\mathbf{n}} \cdot \mathbf{k}$ where $\mathbf{k}$, as always, is the unit vector in the positive z-direction. Thus,

$$ab = \frac{ab'}{\hat{\mathbf{n}} \cdot \mathbf{k}}.$$

Since the area ΔS_l can be approximated with arbitrary accuracy by such rectangles, it follows that

$$\Delta S_l = \frac{\Delta R_l}{\hat{\mathbf{n}}_l \cdot \mathbf{k}},$$

where, of course, $\hat{\mathbf{n}}_l$ is the unit vector normal to the lth plane surface.

We can now rewrite the definition of the surface integral [Equation (II–8)] as

$$
\iint_S G(x,\ y,\ z)\ dS = \lim_{\substack{N\to\infty \\ \text{each}\ \Delta R_l \to 0}} \sum_{l=1}^N G(x_l,\ y_l,\ z_l)\ \frac{\Delta R_l}{\hat{\mathbf{n}}_l \cdot \mathbf{k}}, \quad \text{(II–9)}
$$

where the statement "each $\Delta S_l \to 0$" has been replaced by the equivalent but now more appropriate "each $\Delta R_l \to 0$." We are now obviously well on the road to rewriting the surface integral over S as a double integral over R. In fact,

$$
\lim_{\substack{N\to\infty \\ \text{each}\ \Delta R_l \to 0}} \sum_{l=1}^N \frac{G(x_l,\ y_l,\ z_l)}{\hat{\mathbf{n}}_l \cdot \mathbf{k}}\ \Delta R_l
$$

$$
\equiv \iint_R \frac{G(x,\ y,\ z)}{\hat{\mathbf{n}}(x,\ y,\ z) \cdot \mathbf{k}}\ dx\ dy, \quad \text{(II–10)}
$$

where $\hat{\mathbf{n}}(x,\ y,\ z)$ is the unit vector normal to the surface S at the point $(x,\ y,\ z)$. This is a double integral over R even though it does not quite look like one. What appears to spoil it is that nasty z in G and $\hat{\mathbf{n}}$; a double integral over a region in the xy-plane clearly has no business containing any z's. But the z-dependence is spurious because $(x,\ y,\ z)$ are the coordinates of a point on S, and so $z = f(x,\ y)$. Hence, at the expense of making the integral look even fiercer than it already does in Equation (II–10), we can eliminate the apparent z-dependence of the integrand and write

$$
\iint_R \frac{G[x,\ y,\ f(x,\ y)]}{\hat{\mathbf{n}}[x,\ y,\ f(x,\ y)] \cdot \mathbf{k}}\ dx\ dy. \quad \text{(II–11)}
$$

The faint of heart can take courage; in most cases this integrand reduces quickly to something much simpler and pleasanter looking—a fact we will demonstrate by example below. At this point we introduce the expression for the unit normal vector [Equation (II–4)]. We find

$$
\hat{\mathbf{n}} \cdot \mathbf{k} = \frac{1}{\sqrt{1 + (\partial f/\partial x)^2 + (\partial f/\partial y)^2}},
$$

and so Equation (II–11) becomes

Evaluating
Surface Integrals

$$\iint_S G(x, y, z)\, dS = \iint_R G[x, y, f(x, y)]$$

$$\cdot \sqrt{1 + \left(\frac{\partial f}{\partial x}\right)^2 + \left(\frac{\partial f}{\partial y}\right)^2}\, dx\, dy. \quad \text{(II–12)}$$

Thus, the surface integral of $G(x, y, z)$ over the surface S has been expressed as a double integral of a messy looking function over the region R, the projection of S in the xy-plane. As we remarked above, in practice the integral is usually much less ghastly than it appears written out in Equations (II–11) or (II–12). You will see this in the example we now give.

Let's compute the surface integral

$$\iint_S z^2\, dS,$$

where S is the octant of the sphere of radius 1 centered at the origin as shown in Figure II–15. The projection of S on the xy-

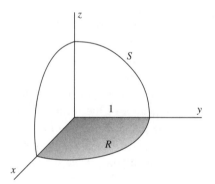

Figure II–15

plane (that is, R) is the area enclosed by the quarter circle. The equation of S is $x^2 + y^2 + z^2 = 1$, or

$$z = f(x, y) = +\sqrt{1 - x^2 - y^2}\,.$$

It follows then that

$$\frac{\partial f}{\partial x} = -\frac{x}{z} \quad \text{and} \quad \frac{\partial f}{\partial y} = -\frac{y}{z},$$

25   so that

$$\sqrt{1 + \left(\frac{\partial f}{\partial x}\right)^2 + \left(\frac{\partial f}{\partial y}\right)^2} = \sqrt{1 + \frac{x^2}{z^2} + \frac{y^2}{z^2}}$$

$$= \frac{1}{z}\sqrt{x^2 + y^2 + z^2} = \frac{1}{z},$$

where we have used $x^2 + y^2 + z^2 = 1$. Hence,

$$\iint_S z^2 \, dS = \iint_R z^2 \frac{1}{z} \, dx \, dy = \iint_R z \, dx \, dy.$$

Substituting for z in terms of x and y, we get

$$\iint_S z^2 \, dS = \iint_R \sqrt{1 - x^2 - y^2} \, dx \, dy.$$

This is an ordinary double integral, and you should verify that its value is $\pi/6$. [*Suggestion:* Convert to polar coordinates: $x = r \cos \theta$, and $y = r \sin \theta$. The integration is then trivial.]

It should be emphasized that the foregoing discussion was based on the assumption that the surface S is described by an equation of the form $z = f(x, y)$; in such a situation a surface integral is converted into a double integral over a region in the xy-plane. But it may happen that a given surface is more conveniently described by an equation of the form $y = g(x, z)$ as in Figure II–16(a). If this is so, then

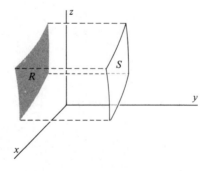

Figure II–16(a)

Evaluating Surface Integrals

$$\iint_S G(x, y, z) \, dS$$

$$= \iint_R G[x, g(x, z), z] \sqrt{1 + \left(\frac{\partial g}{\partial x}\right)^2 + \left(\frac{\partial g}{\partial z}\right)^2} \, dx \, dz,$$

where R is a region in the xz-plane. Similarly, if we have a surface described by $x = h(y, z)$, as in Figure II–16(b), then we use

$$\iint_S G(x, y, z) \, dS$$

$$= \iint_R G[h(y, z), y, z] \sqrt{1 + \left(\frac{\partial h}{\partial y}\right)^2 + \left(\frac{\partial h}{\partial z}\right)^2} \, dy \, dz,$$

where R in this case is a region in the yz-plane. Finally, a surface

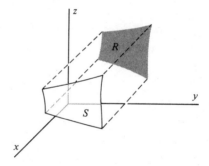

Figure II–16(b)

may have several parts, and it may then be convenient to project different parts on different coordinate planes.

To evaluate surface integrals of the form (II–5), that is,

$$\iint_S \mathbf{F} \cdot \hat{\mathbf{n}} \, dS,$$

we merely replace G by $\mathbf{F} \cdot \hat{\mathbf{n}}$ in Equation (II–12) to get

$$\iint_S \mathbf{F} \cdot \hat{\mathbf{n}} \, dS = \iint_R \mathbf{F} \cdot \hat{\mathbf{n}} \sqrt{1 + \left(\frac{\partial f}{\partial x}\right)^2 + \left(\frac{\partial f}{\partial y}\right)^2} \, dx \, dy.$$

If we now use Equation (II–4) to write this out in detail, we find that the square root factor cancels and we get

Surface Integrals $$\iint_S \mathbf{F} \cdot \hat{\mathbf{n}} \, dS = \iint_R \left\{ -F_x[x, y, f(x, y)] \frac{\partial f}{\partial x} \right.$$
and the
Divergence

$$\left. - F_y[x, y, f(x, y)] \frac{\partial f}{\partial y} + F_z[x, y, f(x, y)] \right\} dx \, dy. \quad \text{(II–13)}$$

We leave it to the reader to write down the analogous formulas when the surface S is given by $y = g(x, z)$ or $x = h(y, z)$, which must be projected onto regions in the xz- and yz-planes, respectively.

This last equation [Equation (II–13)] is enough to make strong men weep, but, as before, in most calculations it quickly reduces to something quite tame. For example, suppose we wish to calculate $\iint_S \mathbf{F} \cdot \hat{\mathbf{n}} \, dS$ where $\mathbf{F}(x, y, z) = \mathbf{i}z - \mathbf{j}y + \mathbf{k}x$ and S is the portion of the plane

$$x + 2y + 2z = 2$$

bounded by the coordinate planes, that is, the triangle reclining gracefully in Figure II–17(a). The normal vector $\hat{\mathbf{n}}$ is chosen so

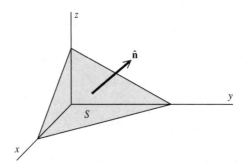

Figure II–17(a)

that it points away from the origin as shown in Figure II–17(a), and we'll project S onto the xy-plane. We have

$$z = f(x, y) = 1 - \frac{x}{2} - y,$$

and so

$$\frac{\partial f}{\partial x} = -\tfrac{1}{2}, \qquad \frac{\partial f}{\partial y} = -1.$$

We also have

$$F_x = z = 1 - \frac{x}{2} - y, \qquad F_y = -y, \qquad F_z = x.$$

Hence

$$\iint_S \mathbf{F} \cdot \hat{\mathbf{n}} \, dS$$

$$= \iint_R \left\{ \left[- \left(1 - \frac{x}{2} - y \right) \right] (-\tfrac{1}{2}) + y(-1) + x \right\} dx \, dy$$

$$= \iint_R \left(\frac{3x}{4} - \frac{3y}{2} + \frac{1}{2} \right) dx \, dy.$$

The region R over which the integral must be taken is shown in Figure II–17(b). The problem has thus been reduced to the com-

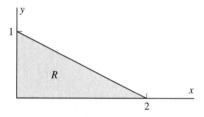

Figure II–17(b)

putation of a rather simple double integral, and you should carry out the integration yourself (the answer is $\tfrac{1}{2}$).

Flux

An integral of the type

$$\iint_S \mathbf{F}(x, y, z) \cdot \hat{\mathbf{n}} \, dS \qquad\qquad \text{(II–14)}$$

is sometimes called the "flux of $\mathbf{F}$." Thus Gauss' law [Equation (II–1)] states that the flux of the electrostatic field over some closed surface is the enclosed charge divided by ϵ_0.

It is useful in obtaining a geometrical feeling for some aspects of vector calculus to understand the significance of the word flux (Latin for "flow") used in this context. For this purpose let us consider a fluid of density ρ moving with velocity **v**. We ask for the total mass of fluid which crosses an area ΔS perpendicular to the direction of flow in a time Δt. Clearly all the fluid in the cylinder of length $v\,\Delta t$ with the patch ΔS as base will cross ΔS in the interval Δt (Figure II–18). The volume of this cylinder is

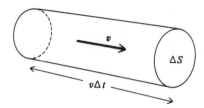

Figure II–18

$v\,\Delta t\,\Delta S$, and it contains a total mass $\rho v\,\Delta t\,\Delta S$. Dividing out the Δt will give the rate of flow. Thus,

$$\left(\begin{array}{c}\text{Rate of flow}\\ \text{through }\Delta S\end{array}\right) = \rho v\,\Delta S.$$

Now let us consider a somewhat more complicated case in which the area ΔS is not perpendicular to the direction of flow (Figure II–19). The volume containing the material which will

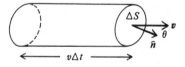

Figure II–19

flow through ΔS in time Δt is now just the volume of the little skewed cylinder shown in the diagram. The volume is $v\,\Delta t\,\Delta S \cos\theta$, where θ is the angle between the velocity vector **v** and $\hat{\mathbf{n}}$, the unit vector normal to ΔS and pointing outward from the skewed cylinder. But $v\cos\theta = \mathbf{v}\cdot\hat{\mathbf{n}}$. So, multiplying by ρ and dividing by Δt, we get

$$\left(\begin{array}{c} \text{Rate of flow} \\ \text{through } \Delta S \end{array}\right) = \rho \mathbf{v} \cdot \hat{\mathbf{n}} \, \Delta S.$$

Finally, consider a surface S in some region of space containing flowing matter (Figure II–20). Approximate the surface by a

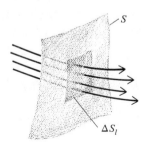

Figure II–20

polyhedron. By the above argument, the rate at which matter flows through the lth face of this polyhedron is approximately

$$\rho(x_l, y_l, z_l)\mathbf{v}(x_l, y_l, z_l) \cdot \hat{\mathbf{n}}_l \, \Delta S_l.$$

Here, of course, (x_l, y_l, z_l) are the coordinates of the point on the lth face at which it is tangent to S, and $\hat{\mathbf{n}}_l$ is the unit vector normal to the lth face. Summing over all the faces and taking the limit, we get

$$\left(\begin{array}{c} \text{Rate of flow} \\ \text{through } S \end{array}\right) = \iint_S \rho(x, y, z)\mathbf{v}(x, y, z) \cdot \hat{\mathbf{n}} \, dS.$$

If S happens to be a closed surface and there is a net rate of flow *out* of the volume it encloses, then you can convince yourself that this integral will be positive, and if there is a net rate of flow *in*, the integral will be negative.

If in this last equation we put

$$\mathbf{F}(x, y, z) = \rho(x, y, z)\mathbf{v}(x, y, z),$$

the integral is seen to be formally identical with that in Equation (II–14). For this reason any integral of the form (II–14) is called "the flux of $\mathbf{F}$ over the surface S," even when the function $\mathbf{F}$ is *not* the product of a density and a velocity! The reason for stressing this point about flux is that, misnomer though it may be, it nonetheless gives a good geometric or physical picture of Gauss' law: The electric field "flows" out of a surface enclosing charge, and the "amount" of this "flow" is proportional to the net charge enclosed. *Warning:* This is not to be taken literally; the electric field is not flowing in the sense in which fluid flows. It is merely picturesque language intended to aid us in understanding the physics in Gauss' law.

Using Gauss' Law to Find the Field

Having rejected the two expressions for **E** [Equations (I–4) and (I–7)], we find that the only candidate left for providing us with a good general method for calculating the field is Gauss' law. At first glance it does not appear to be a very likely candidate because, unlike Equations (I–4) and (I–7), it is not an *explicit* expression for **E**. That is, it does not say "**E** equals something." Rather, it says "The flux of **E** (the surface integral of the normal component of **E**) equals something." Thus, to use Gauss' law, we must "disentangle" **E** from its surroundings. Despite this, there are situations in which Gauss' law can be used to find the field as an example will now show.

Consider a point charge q placed at the origin of a coordinate system. Symmetry considerations tell us two things about its electric field: (1) It must be in the radial direction (that is, it must point directly toward, or directly away from, the origin), and (2) it must have the same magnitude at all points on the surface of a sphere centered at the origin. Stating this in symbols, we have $\mathbf{E} = \hat{\mathbf{e}}_r E(r)$, where $\hat{\mathbf{e}}_r = \mathbf{r}/r$ is a unit vector in the radial direction. Thus, Gauss' law becomes

$$\iint_S E(r)\hat{\mathbf{e}}_r \cdot \hat{\mathbf{n}}\ dS = q/\epsilon_0.$$

If, for the surface S, we now choose a spherical shell of radius r centered at the origin, a little thought will convince you that $\hat{\mathbf{n}} = \hat{\mathbf{e}}_r$, so that $\hat{\mathbf{n}} \cdot \hat{\mathbf{e}}_r = 1$ and we get

$$\iint_S E(r)\ dS = q/\epsilon_0.$$

This integral is trivial to perform if we recognize that r is a constant over the spherical surface S. This means that $E(r)$ is also a constant on S and we get[4]

$$\iint_S E(r)\, dS = E(r) \iint_S dS = 4\pi r^2 E(r) = q/\epsilon_0,$$

whence

$$E(r) = \frac{1}{4\pi\epsilon_0} \frac{q}{r^2}$$

and

$$\mathbf{E(r)} = \hat{\mathbf{e}}_r E(r) = \frac{\hat{\mathbf{e}}_r}{4\pi\epsilon_0} \frac{q}{r^2},$$

in agreement with Equation (I–2).

We can see from this example how heavily we depend on symmetry when using Gauss' law to obtain the field. In fact, to use Gauss' law in the form given in Equation (II–1) requires even more symmetry and simplicity than Equations (I–4) and (I–7). The blunt truth is that this form of the law yields the electric field in a grand total of three situations (and combinations thereof): (1) a spherically symmetric distribution of charge (of which the point charge considered above is a special case), (2) an infinitely long cylindrically symmetric distribution (including the case of an infinitely long uniformly distributed line of charge), and (3) an infinite slab of charge (including as a special case an infinite uniformly charged plane).[5] The real value of Equation (II–1) is that it can be twisted and beaten into a more useful form.

What is it about Equation (II–1) that makes it difficult to find **E**? To answer this question, suppose we are doing a numerical calculation on a computer and wish to evaluate $\iint_S \mathbf{E} \cdot \hat{\mathbf{n}}\, dS$. The standard procedure for dealing with integrals numerically is to approximate them as sums, a rather obvious thing to do since an integral, after all, is the *limit* of a sum. Thus, suppose we divide the surface S into, say, 10 patches. We then have as an approximation to Equation (II–1)

---

[4] Shortcuts like this often make it possible to evaluate surface integrals without using all the paraphernalia we discussed above. Further examples are given in Problem II–10.

[5] Examples of these are given in Problems II–11, II–12, and II–13.

$$\sum_{i=1}^{10} \mathbf{E}_l \cdot \hat{\mathbf{n}}_l \, \Delta S_l \simeq q/\epsilon_0,$$

where $\mathbf{E}_l$ is the value of $\mathbf{E}$, and $\hat{\mathbf{n}}_l$ is the unit normal, somewhere on the *l*th patch. There is little or no hope of finding $\mathbf{E}$ from this: it is one equation in the 10 unknowns $\mathbf{E}_1, \mathbf{E}_2, \ldots, \mathbf{E}_{10}$. Furthermore, it is probably not very accurate. To improve the accuracy, we might make 100 subdivisions rather than just 10 to get

$$\sum_{l=1}^{100} \mathbf{E}_l \cdot \hat{\mathbf{n}}_l \, \Delta S_l \simeq q/\epsilon_0.$$

Much more accurate! And much more hopeless, too, because this is one equation in 100 unknowns. Even more accurate (and more hopeless) is

$$\iint_S \mathbf{E} \cdot \hat{\mathbf{n}} \, dS = q/\epsilon_0,$$

which is one equation in *infinitely many* unknowns. These unknowns are, of course, the values of $\mathbf{E} \cdot \hat{\mathbf{n}}$ at every one of the infinitely many points of the surface S.[6]

We have now isolated the trouble with Equation (II–1): it involves an entire surface and therefore the value of $\mathbf{E} \cdot \hat{\mathbf{n}}$ at infinitely many points. If, somehow, we could deal with the "flux at a single point" (whatever *that* may mean!) rather than the flux through a surface, perhaps then Gauss' law might yield something tractable. How might we arrange this? For simplicity let us surround some point P by a set of concentric spherical shells S_1, S_2, S_3, and so on (Figure II–21), and calculate the flux Φ_1, Φ_2, Φ_3, and so on, through each shell. We might then attempt to define the "flux at the point P" as the limiting value approached by the sequence of fluxes calculated this way over smaller and smaller shells centered at P.

This sounds good; it has a heartening "mathematical" ring to it. Unfortunately, it does not work because (assuming the charge density is finite everywhere) the sequence of fluxes, calculated as described above, approaches zero for *any* point P. This is fairly obvious since it is merely the statement that the flux through a

---

[6] The reason Gauss' law yields the expression for the field of a point charge examined above is that symmetry in that case shows the infinitely many unknowns are all equal. This turns Gauss' law into one equation in *one* unknown.

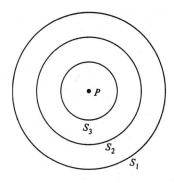

Figure II–21

surface tends to zero as the surface shrinks to a point. Since our objective was to find a way to determine the flux at a point, and thereby learn something about the field at that point, and since we get zero at *any* point no matter what the field there may be, we have obviously not obtained what we want.

It is useful to give a physicist's rough-and-ready proof of the fact that the flux goes to zero as the surface shrinks down to a point, for even though this fact may be obvious, the proof will suggest how to pull this chestnut out of the fire. For this purpose we note that if $\bar{\rho}_{\Delta V}$ denotes the average density of electric charge [Equation I–5] in some region of volume ΔV, then the total charge in ΔV is $\bar{\rho}_{\Delta V} \Delta V$. Thus Gauss' law [Equation (II–1)] may be written

$$\iint_S \mathbf{E} \cdot \hat{\mathbf{n}} \, dS = \bar{\rho}_{\Delta V} \, \Delta V / \epsilon_0, \qquad \text{(II–15)}$$

where, as indicated in Figure II–22, the surface integral is taken

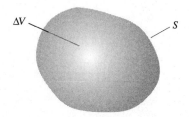

Figure II–22

over the surface S which encloses the volume ΔV. From this expression [Equation (II–15)] we can see the validity of our assertion: As $S \to 0$, the enclosed volume ΔV must, of course, also approach zero. Thus, the flux also tends to zero and the assertion is proved.[7] We have not only given a proof, but (and this is the point) we can now isolate a quantity which does *not* vanish as $S \to 0$. Dividing Equation (II–15) by ΔV, we get

$$\frac{1}{\Delta V} \iint_S \mathbf{E} \cdot \hat{\mathbf{n}} \, dS = \bar{\rho}_{\Delta V}/\epsilon_0.$$

This expression, awkward and unappealing though it may be, is nonetheless close to what we are after, even though it still involves an integral of $\mathbf{E}$ over an entire surface. For if we *now* take the limit as S shrinks to zero about some point in ΔV whose coordinates are (x, y, z), then, as we see from Equation (I–6), the average density $\bar{\rho}_{\Delta V}$ approaches $\rho(x, y, z)$, the density at (x, y, z), and we get

$$\lim_{\substack{\Delta V \to 0 \\ \text{about } (x,y,z)}} \frac{1}{\Delta V} \iint_S \mathbf{E} \cdot \hat{\mathbf{n}} \, dS = \rho(x, y, z)/\epsilon_0. \qquad \text{(II–16)}$$

This expression is admittedly downright hideous and whether it will be of any practical use whatever depends on our being able to pound the left-hand side into a form which looks and acts at least half-civilized. We turn to this task now.

The Divergence

Let us consider the surface integral of some arbitrary vector function $\mathbf{F}(x, y, z)$:

$$\iint_S \mathbf{F} \cdot \hat{\mathbf{n}} \, dS.$$

We shall be interested in the ratio of this integral to the volume enclosed by the surface S as the volume shrinks to zero about some point, for that is exactly the type of quantity which appears in Equation (II–16). This limit is important enough to warrant a special name and notation. It is called the *divergence of* $\mathbf{F}$ and is designated div $\mathbf{F}$. Thus,

---

[7] This line of reasoning and the conclusion must be altered if the system contains point charges.

$$\text{div } \mathbf{F} \equiv \lim_{\substack{\Delta V \to 0 \\ \text{about } (x,y,z)}} \frac{1}{\Delta V} \iint_S \mathbf{F} \cdot \hat{\mathbf{n}} \, dS. \qquad \text{(II–17)}$$

This quantity is clearly a scalar. Furthermore, it will, in general, have different values at different points (x, y, z). Thus the divergence of a vector function is a *scalar function* of position.

Equation (II–16) can now be written

$$\text{div } \mathbf{E} = \rho/\epsilon_0. \qquad \text{(II–18)}$$

At this stage, however, our fancy new notation has only a cosmetic value, helping to beautify an ugly equation. Whether it has any practical value as well is the matter taken up in the following discussion in which we actually calculate the limit of the ratio of flux to enclosed volume and find that it can be expressed reasonably simply in terms of certain partial derivatives. Before turning to this calculation, however, it's worth mentioning that if we take our new terminology literally, we can interpret Equation (II–18) to mean that the field "diverges" from a point, and how much it diverges, so to speak, depends on how much charge there is at that point as represented by the density there.

Our next order of business is to find the reasonably simple expression for the divergence of a vector function promised above. Thus, consider a small rectangular parallelepiped[8] with edges of length Δx, Δy, and Δz parallel to the coordinate axes (Figure II–23). Let the point at the center of the little cuboid have

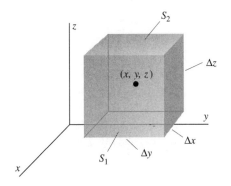

Figure II–23

---

[8] Henceforth we'll refer to this as a "cuboid," a made-up term that takes less time and space than the sesquipedalian "rectangular parallelepiped."

coordinates (x, y, z). We calculate the surface integral of **F** over the surface of the cuboid by regarding the integral as a sum of six terms, one for each cuboid face. We begin by considering the face marked S_1 in the figure. We want

$$\iint_{S_1} \mathbf{F} \cdot \hat{\mathbf{n}} \ dS.$$

Now it is clear the unit vector normal to this face and pointing outward from the enclosed volume is **i**. Thus, since $\mathbf{F} \cdot \mathbf{i} = F_x$, the above integral is

$$\iint_{S_1} F_x(x, y, z) \ dS.$$

By assumption the cuboid is small (eventually we shall take the limit as it shrinks to zero). We can therefore calculate this integral approximately as F_x evaluated at the center of the face S_1 multiplied by the area of the face.[9] The coordinates of the center of S_1 are $(x + \Delta x/2, y, z)$. Thus,

$$\iint_{S_1} F_x(x, y, z) \ dS \simeq F_x\left(x + \frac{\Delta x}{2}, y, z\right) \Delta y \ \Delta z. \qquad \text{(II–19)}$$

The same kind of reasoning applied to the opposite face S_2 [whose outward normal is $-\mathbf{i}$ and whose center is at $(x - \Delta x/2, y, z)$] leads to

$$\iint_{S_2} \mathbf{F} \cdot \hat{\mathbf{n}} \ dS = -\iint_{S_2} F_x \ dS$$

$$\simeq -F_x\left(x - \frac{\Delta x}{2}, y, z\right) \Delta y \ \Delta z. \qquad \text{(II–20)}$$

Adding together the contributions of these two faces [Equations (II–19) and (II–20)], we get

[9] The rationale behind this is as follows: There is a mean value theorem, which tells us that the integral of F_x over S_1 is equal to the area of S_1 multiplied by the function evaluated *somewhere* on S_1. Since S_1 is small, the point where we *should* evaluate F_x and the point where we *do* evaluate it (that is, the center) must be close together, and F_x must have nearly the same value at the two points. Hence what our procedure gives us is a good approximation to the value of the integral. Furthermore, as the cuboid shrinks to zero, the two points get closer and closer so that in the limit our result [Equation (II–22)] will be exact.

$$\iint_{S_1+S_2} \mathbf{F} \cdot \hat{\mathbf{n}}\, dS$$

$$= \left[F_x\left(x + \frac{\Delta x}{2}, y, z \right) - F_x\left(x - \frac{\Delta x}{2}, y, z \right) \right] \Delta y\, \Delta z$$

$$= \frac{F_x\left(x + \dfrac{\Delta x}{2}, y, z \right) - F_x\left(x - \dfrac{\Delta x}{2}, y, z \right)}{\Delta x} \, \Delta x\, \Delta y\, \Delta z.$$

Recognizing that $\Delta x\, \Delta y\, \Delta z = \Delta V$, the volume of the cuboid, we have

$$\frac{1}{\Delta V} \iint_{S_1+S_2} \mathbf{F} \cdot \hat{\mathbf{n}}\, dS$$

$$= \frac{F_x\left(x + \dfrac{\Delta x}{2}, y, z \right) - F_x\left(x - \dfrac{\Delta x}{2}, y, z \right)}{\Delta x}. \qquad \text{(II–21)}$$

We now must take the limit of this as ΔV approaches zero.[10] But, of course, as ΔV goes to zero, so do each of the sides of the cuboid. Thus, on the right-hand side of Equation (II–21) we can write $\lim_{\Delta x \to 0}$ in place of $\lim_{\Delta V \to 0}$, and we find

$$\lim_{\Delta V \to 0} \frac{1}{\Delta V} \iint_{S_1+S_2} \mathbf{F} \cdot \hat{\mathbf{n}}\, dS$$

$$= \lim_{\Delta x \to 0} \frac{F_x\left(x + \dfrac{\Delta x}{2}, y, z \right) - F_x\left(x - \dfrac{\Delta x}{2}, y, z \right)}{\Delta x} = \frac{\partial F_x}{\partial x}$$

evaluated at (x, y, z). This last equality follows from the definition of the partial derivative. It should come as no surprise that the other two pairs of faces of the cuboid contribute $\partial F_y/\partial y$ and $\partial F_z/\partial z$. Thus,

$$\lim_{\Delta V \to 0} \frac{1}{\Delta V} \iint_S \mathbf{F} \cdot \hat{\mathbf{n}}\, dS = \frac{\partial F_x}{\partial x} + \frac{\partial F_y}{\partial y} + \frac{\partial F_z}{\partial z}.$$

[10] Note that we have postponed calculating the contributions from the other four faces of the cuboid.

The limit on the left-hand side of this last equation, as we have already remarked, is the divergence of **F** [Equation (II–17)]. Thus we have just demonstrated that

$$\text{div } \mathbf{F} = \frac{\partial F_x}{\partial x} + \frac{\partial F_y}{\partial y} + \frac{\partial F_z}{\partial z}. \tag{II–22}$$

It can be shown that this result is independent of the shape of the volume used to obtain it (see Problem II–17).

Using Equation (II–22) to find the divergence of a vector function is a straightforward matter, but we'll give an example just for the record. Consider the function

$$\mathbf{F}(x, y, z) = \mathbf{i}x^2 + \mathbf{j}xy + \mathbf{k}yz.$$

We have

$$\frac{\partial F_x}{\partial x} = 2x, \qquad \frac{\partial F_y}{\partial y} = x, \qquad \text{and} \qquad \frac{\partial F_z}{\partial z} = y.$$

Thus,

$$\text{div } \mathbf{F} = 2x + x + y = 3x + y.$$

Returning now to the electrostatic field, we combine Equations (II–18) and (II–22) to get

$$\frac{\partial E_x}{\partial x} + \frac{\partial E_y}{\partial y} + \frac{\partial E_z}{\partial z} = \rho/\epsilon_0. \tag{II–23}$$

This equation, which is much more general than our derivation of it suggests, is one of Maxwell's equations and is completely equivalent to Gauss' law [Equation (II–1)]. It is sometimes called the "differential form" of Gauss' law.

We have now arrived at our goal (almost!) for we have related a property of the electrostatic field at a point (that is, its divergence) to a known quantity (the charge density) at that point. In all fairness it should be said that Equation (II–23) can in a sense be regarded as a single (differential) equation in *three* unknowns. (E_x, E_y, E_z) and for this reason is not often used in this form to find the field. It turns out, however, that the three components of

E can be related to each other very elegantly; when we develop that relationship, we shall return to this question of finding a convenient means of calculating **E**.

The Divergence in Cylindrical and Spherical Coordinates

One often sees Equation (II–22) given as the definition of the divergence of the vector function **F**. While this is certainly acceptable, we much prefer to define the divergence as the limit of flux to volume as stated in Equation (II–16). Equation (II–22) is then merely the form the divergence takes in Cartesian coordinates. In other coordinate systems it looks quite different. For example, in cylindrical coordinates the function **F** has three components, which you will not be shocked to learn are designated F_r, F_θ, and F_z [see Figure 11–24(a)]. To obtain the divergence

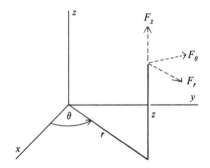

Figure II–24(a)

of **F** in cylindrical coordinates, we consider the "cylindrical cuboid" shown in Figure II–24(b) with volume $\Delta V = r\,\Delta r\,\Delta\theta\,\Delta z$

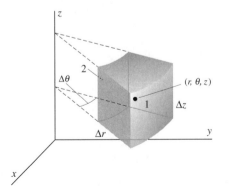

41   Figure II–24(b)

and center at the point (r, θ, z).[11] The flux of **F** through the face marked "1" is

$$\iint_{S_1} \mathbf{F} \cdot \hat{\mathbf{n}} \, dS = \iint_{S_1} F_r \, dS$$

$$\simeq F_r\left(r + \frac{\Delta r}{2}, \theta, z\right)\left(r + \frac{\Delta r}{2}\right) \Delta\theta \, \Delta z,$$

while through the face marked "2" it is

$$\iint_{S_2} \mathbf{F} \cdot \hat{\mathbf{n}} \, dS = -\iint_{S_2} F_r \, dS$$

$$\simeq -F_r\left(r - \frac{\Delta r}{2}, \theta, z\right)\left(r - \frac{\Delta r}{2}\right) \Delta\theta \, \Delta z.$$

Adding these two results and dividing by the volume ΔV of the cuboid, we find

$$\frac{1}{\Delta V} \iint_{S_1 + S_2} \mathbf{F} \cdot \hat{\mathbf{n}} \, dS$$

$$\simeq \frac{1}{r\Delta r}\left[\left(r + \frac{\Delta r}{2}\right) F_r\left(r + \frac{\Delta r}{2}, \theta, z\right)\right.$$

$$\left. - \left(r - \frac{\Delta r}{2}\right) F_r\left(r - \frac{\Delta r}{2}, \theta, z\right)\right],$$

which in the limit as Δr (and therefore ΔV) approaches zero becomes

$$\frac{1}{r} \frac{\partial}{\partial r} (r F_r).$$

[11] Note that in the Cartesian case (Figure II–23) each face of the cuboid is given by an equation of the form $x =$ constant, $y =$ constant, or $z =$ constant. In the same way each face of the surface in Figure II–24(b) is given by an equation of the form $r =$ constant, $\theta =$ constant, or $z =$ constant.

The Del Notation   Arguing in an analogous way for the other four faces (see Problem II–18), we arrive finally at the expression for the divergence in cylindrical coordinates:

$$\text{div } \mathbf{F} = \frac{1}{r}\frac{\partial}{\partial r}(rF_r) + \frac{1}{r}\frac{\partial F_\theta}{\partial \theta} + \frac{\partial F_z}{\partial z}. \tag{II–24}$$

In spherical coordinates where the components of $\mathbf{F}$ are F_r, F_θ, and F_ϕ (see Figure II–25) similar reasoning (see Problem

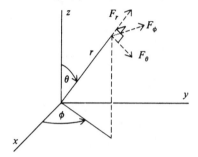

Figure II–25

II–21) leads to the expression

$$\text{div } \mathbf{F} = \frac{1}{r^2}\frac{\partial}{\partial r}(r^2 F_r) + \frac{1}{r\sin\theta}\frac{\partial}{\partial \theta}(\sin\theta\, F_\theta) + \frac{1}{r\sin\theta}\frac{\partial F_\phi}{\partial \phi}. \tag{II–25}$$

The Del Notation

There is a special notation in terms of which the divergence may be written. There would be little or no reason for introducing it if it served only to provide another way of writing "div," but as we shall soon see, it has considerable usefulness in vector calculus.

Let us define a quantity designated ∇ (read "del") by the following rather peculiar looking equation:

$$\nabla = \mathbf{i}\frac{\partial}{\partial x} + \mathbf{j}\frac{\partial}{\partial y} + \mathbf{k}\frac{\partial}{\partial z}.$$

If we take the dot product of ∇ and some vector function $\mathbf{F} = \mathbf{i}F_x + \mathbf{j}F_y + \mathbf{k}F_z$, we get

$$\nabla \cdot \mathbf{F} = \left(\mathbf{i} \frac{\partial}{\partial x} + \mathbf{j} \frac{\partial}{\partial y} + \mathbf{k} \frac{\partial}{\partial z} \right) \cdot (\mathbf{i}F_x + \mathbf{j}F_y + \mathbf{k}F_z)$$

$$= \frac{\partial}{\partial x} F_x + \frac{\partial}{\partial y} F_y + \frac{\partial}{\partial z} F_z.$$

Now we interpret the "product" of $\partial/\partial x$ and F_x as a partial derivative; that is,

$$\frac{\partial}{\partial x} F_x \equiv \frac{\partial F_x}{\partial x}.$$

There are similar equations for the two other "products" $(\partial/\partial y)F_y$ and $(\partial/\partial z)F_z$. With this convention we recognize $\nabla \cdot \mathbf{F}$ ("del dot $\mathbf{F}$") as the same as div $\mathbf{F}$, and henceforth, to conform with modern notational practice, we shall always use $\nabla \cdot \mathbf{F}$ to indicate the divergence. Thus, Equations (II–18) and (II–23) will be written

$$\nabla \cdot \mathbf{E} = \rho/\epsilon_0.$$

Mathematicians call a symbol like ∇ an operator. When we "operate" with ∇ by dotting it into a vector function we get the divergence of that function, as we have just seen. In subsequent discussions we shall introduce three other quantities (gradient, curl, and Laplacian) all of which are operators and all of which can be written in terms of ∇.

The Divergence Theorem

For the remainder of this chapter we digress from the mainstream of our narrative to discuss a famous theorem which asserts a remarkable connection between surface integrals and volume integrals. Although this relation may be suggested by the work we have done in electrostatics, the theorem is a mathematical statement holding under quite general circumstances. It is independent of any physics and is applicable in many different places.

It is called the divergence theorem and sometimes Gauss' theorem (not to be confused with Gauss' *law*).

We shall not give a mathematically rigorous proof of the divergence theorem; such a proof is given in many texts in advanced calculus. Instead we present here another physicist's rough-and-ready proof. Thus, consider a closed surface S. Subdivide the volume V enclosed by S arbitrarily into N subvolumes, one of which is shown in Figure II–26 (drawn as a cube for conven-

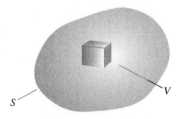

Figure II–26

ience). We begin our proof by asserting that the flux of an arbitrary vector function $\mathbf{F}(x, y, z)$ through the surface S equals the sum of the fluxes through the surfaces of each of the subvolumes:

$$\iint_S \mathbf{F} \cdot \hat{\mathbf{n}} \, dS = \sum_{l=1}^{N} \iint_{S_l} \mathbf{F} \cdot \hat{\mathbf{n}} \, dS. \qquad \text{(II–26)}$$

Here S_l is the surface which encloses the subvolume ΔV_l. To establish Equation (II–26), consider two adjacent subvolumes (Figure II–27). Let their common face be denoted S_0. The flux

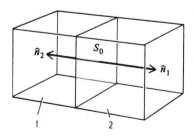

45 Figure II–27

through the subvolume marked ''1'' in Figure II–27 includes, of course, a contribution from S_0, which is

$$\iint_{S_0} \mathbf{F} \cdot \hat{\mathbf{n}}_1 \, dS.$$

Here $\hat{\mathbf{n}}_1$ is a unit vector normal to the face S_0, and by our usual convention, it points outward from subvolume 1. The flux through the subvolume marked ''2'' also includes a contribution from S_0:

$$\iint_{S_0} \mathbf{F} \cdot \hat{\mathbf{n}}_2 \, dS.$$

The vector $\hat{\mathbf{n}}_2$ is a unit normal which points outward from subvolume 2. Clearly $\hat{\mathbf{n}}_1 = -\hat{\mathbf{n}}_2$. Thus, in forming the sum in Equation (II–26), we shall include, among other things, the pair of terms

$$\iint_{S_0} \mathbf{F} \cdot \hat{\mathbf{n}}_1 \, dS + \iint_{S_0} \mathbf{F} \cdot \hat{\mathbf{n}}_2 \, dS =$$

$$\iint_{S_0} \mathbf{F} \cdot \hat{\mathbf{n}}_1 \, dS - \iint_{S_0} \mathbf{F} \cdot \hat{\mathbf{n}}_1 \, dS = 0.$$

We see that these terms cancel each other and there is no net contribution to the sum in Equation (II–26) due to the face S_0. In fact this sort of cancellation will obviously occur for any subvolume surface which is common to two adjacent subvolumes. But *all* subvolume surfaces are common to two adjacent subvolumes except those which are part of the original (''outer'') surface S. Hence the only terms in the sum in Equation (II–26) which survive come from those subvolume surfaces which, taken together, constitute the surface S. This establishes the validity of Equation (II–26).

We now rewrite Equation (II–26) in the following curious fashion:

$$\iint_S \mathbf{F} \cdot \hat{\mathbf{n}} \, dS = \sum_{l=1}^{N} \left[\frac{1}{\Delta V_l} \iint_{S_l} \mathbf{F} \cdot \hat{\mathbf{n}} \, dS \right] \Delta V_l. \quad \text{(II–27)}$$

This clearly alters nothing since we have just multiplied and divided each term of the sum by ΔV_l, the subvolume enclosed by the surface S_l. We can now imagine partitioning the or' ume V into an ever larger number of smaller and sm volumes. In other words, we take the limit of the sum in Equation (II–27) as the number of subdivisions tends to infinity and each ΔV_l tends to zero. We recognize that the limit of the quantity in square brackets in Equation (II–27) is, by definition, $(\nabla \cdot \mathbf{F})_l$, that is, the divergence of $\mathbf{F}$ evaluated at the point about which ΔV_l is shrinking. Thus, for each ΔV_l very small, Equation (II–27) becomes

$$\iint_S \mathbf{F} \cdot \hat{\mathbf{n}} \, dS \simeq \sum_{l=1}^{N} (\nabla \cdot \mathbf{F})_l \, \Delta V_l. \qquad \text{(II–28)}$$

Further, in the limit, this sum is, again by definition, the triple integral of $\nabla \cdot \mathbf{F}$ over the volume enclosed by S:

$$\lim_{\substack{N \to \infty \\ \text{each } \Delta V_l \to 0}} \sum_{l=1}^{N} (\nabla \cdot \mathbf{F})_l \, \Delta V_l \equiv \iiint_V \nabla \cdot \mathbf{F} \, dV. \qquad \text{(II–29)}$$

Putting together Equations (II–26) through (II–29), we arrive at our result:

$$\iint_S \mathbf{F} \cdot \hat{\mathbf{n}} \, dS = \iiint_V \nabla \cdot \mathbf{F} \, dV. \qquad \text{(II–30)}$$

flux

This is the divergence theorem. In words it says that the flux of a vector function through some closed surface equals the triple integral of the divergence of that function over the volume enclosed by the surface.

The major reason the proof given above is not rigorous is that a triple integral is defined as the limit of a sum of the form

$$\sum_l g(x_l, y_l, z_l) \, \Delta V_l,$$

where the function g is well-defined. In Equation (II–27), however, the quantity multiplying the volume element ΔV_l in each term of the sum is not a well-defined function in this sense. That is, as ΔV_l tends to zero the quantity in the square brackets changes; it can be identified as the divergence of $\mathbf{F}$ only in the limit. A careful, rigorous treatment would show that Equation (II–30) is valid if $\mathbf{F}$ (that is, F_x, F_y, and F_z) is continuous and differentiable, and its first derivatives are continuous in V and on S.

Now let's illustrate the divergence theorem. Since endless pages of hideous integrals will not serve our purpose, we'll use a simple example. Let

$$\mathbf{F}(x, y, z) = \mathbf{i}x + \mathbf{j}y + \mathbf{k}z$$

and choose for S the surface shown in Figure II–28, consisting

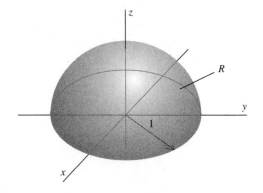

Figure II–28

of the hemispherical shell of radius 1 and the region R of the xy-plane enclosed by the unit circle. On the hemisphere we have $\hat{\mathbf{n}} = \mathbf{i}x + \mathbf{j}y + \mathbf{k}z$, so that $\hat{\mathbf{n}} \cdot \mathbf{F} = x^2 + y^2 + z^2 = 1$. Thus, on the hemisphere,

$$\iint \mathbf{F} \cdot \hat{\mathbf{n}} \, dS = \iint dS = 2\pi,$$

where the last equality follows from the fact that the integral is merely the surface area of the unit hemisphere. On the region R we have $\hat{\mathbf{n}} = -\mathbf{k}$ so that $\hat{\mathbf{n}} \cdot \mathbf{F} = -z$. Hence, on R,

$$\iint \mathbf{F} \cdot \hat{\mathbf{n}} \, dS = -\iint z \, dx \, dy = 0$$

because $z = 0$ everywhere on R. Thus, there is no contribution to the surface integral from the circular region R and

$$\iint_S \mathbf{F} \cdot \hat{\mathbf{n}} \, dS = 2\pi.$$

Next we find by a trivial calculation that $\nabla \cdot \mathbf{F} = 3$. It follows then that

$$\iiint_V \nabla \cdot \mathbf{F} \, dV = 3 \iiint_V dV = 3 \frac{2\pi}{3} = 2\pi,$$

where we use the fact that the volume of the unit hemisphere is $2\pi/3$. Since the surface and volume integrals are equal, this illustrates Equation (II–30).

Two Simple Applications of the Divergence Theorem

As one example of the use of the divergence theorem we give an alternative derivation of Equation (II–18), the analysis of which led us to the divergence theorem itself. In other words, this is how easy it would have been if we had known the divergence theorem to begin with!

We start with Gauss' law in the form

$$\iint_S \mathbf{E} \cdot \hat{\mathbf{n}} \, dS = \frac{1}{\epsilon_0} \iiint_V \rho \, dV. \qquad \text{(II–31)}$$

Next we apply the divergence theorem to the surface integral in the above equation to get

$$\iint_S \mathbf{E} \cdot \hat{\mathbf{n}} \, dS = \iiint_V \nabla \cdot \mathbf{E} \, dV. \qquad \text{(II–32)}$$

Thus, combining Equations (II–31) and II–32), we find

$$\iiint_V \nabla \cdot \mathbf{E} \, dV = \frac{1}{\epsilon_0} \iiint_V \rho \, dV.$$

In general, if two volume integrals are equal, it is not necessarily true that their integrands are equal. It might be that the integrals are equal only over the particular volume of integration V, and by integrating over a different volume, we would wreck the equality. In the present case, however, this is not true because

Gauss' law holds for any arbitrary volume V, and we cannot upset the equality by changing the volume. But this can be so only if the integrands are equal. Hence,

$$\nabla \cdot \mathbf{E} = \rho/\epsilon_0,$$

which should look familiar!

Another example of the use of the divergence theorem is the following. Suppose that in some region of space "stuff" (matter, electric charge, anything) is moving (Figure II–29). Let the den-

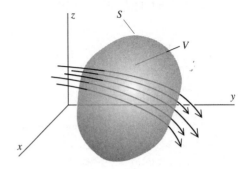

Figure II–29

sity of this stuff at any point (x, y, z) and any time t be $\rho(x, y, z, t)$ and let its velocity be $\mathbf{v}(x, y, z, t)$. Further, suppose this stuff is *conserved;* that is, it is neither created nor destroyed. Concentrating on some arbitrary volume V in space, we ask: What is the rate at which the amount of stuff in this volume is changing? At any time t the amount of stuff in V is

$$\iiint_V \rho(x, y, z, t)\, dV$$

and the rate at which it is changing is

$$\frac{d}{dt} \iiint_V \rho(x, y, z, t)\, dV = \iiint_V \frac{\partial \rho}{\partial t}\, dV.$$

(To be able to move the derivative under the integral sign this way requires that $\partial \rho/\partial t$ be continuous.)

Next we recall from an earlier discussion that the rate at which stuff flows through a surface S is

$$\iint_S \rho \mathbf{v} \cdot \hat{\mathbf{n}} \, dS.$$

We then assert that the rate at which the amount of stuff in V is changing is equal to the rate at which it is flowing through the enclosing surface S; in equation form this statement reads

$$\iiint_V \frac{\partial \rho}{\partial t} \, dV = - \iint_S \rho \mathbf{v} \cdot \hat{\mathbf{n}} \, dS.$$

There are two features about this equation that require discussion:

1. The negative sign must be included because the surface integral as defined is positive for a net flow *out* of the volume, but a net flow out means the amount of stuff in the volume is decreasing.
2. This equation states that the amount of stuff in V can change only as a result of stuff flowing across the boundary S. If stuff were being created or destroyed in V, terms would have to be included in the equation to reflect that fact. The absence of any such terms is thus an expression of the conservation of the stuff.

Now, finally, let us apply the divergence theorem. We find

$$\iint_S \rho \mathbf{v} \cdot \hat{\mathbf{n}} \, dS = \iiint_V \nabla \cdot (\rho \mathbf{v}) \, dV.$$

Hence,

$$\iiint_V \frac{\partial \rho}{\partial t} \, dV = - \iiint_V \nabla \cdot (\rho \mathbf{v}) \, dV.$$

Arguing as we did above that V is an arbitrary volume, we can then say

$$\frac{\partial \rho}{\partial t} = -\nabla \cdot (\rho \mathbf{v}). \tag{II–33}$$

Usually we define the *current density* $\mathbf{J} = \rho\mathbf{v}$ and write Equation (II–31) as

$$\frac{\partial\rho}{\partial t} + \nabla \cdot \mathbf{J} = 0.$$

An equation of this type is referred to as a *continuity equation* and is, as we have seen, an expression of a conservation law (see Problems III–20, III–21, and IV–21). Besides playing an important role in electromagnetic theory, it is a basic equation both in hydrodynamics and diffusion theory. Finally, considerations similar to those which led to the continuity equation are involved in the analysis of heat flow.

PROBLEMS

II–1 Find a unit vector $\hat{\mathbf{n}}$ normal to each of the following surfaces.

(a) $z = 2 - x - y$.   (c) $z = (1 - x^2)^{1/2}$.
(b) $z = (x^2 + y^2)^{1/2}$.   (d) $z = x^2 + y^2$.
          (e) $z = (1 - x^2/a^2 - y^2/a^2)^{1/2}$.

II–2 (a) Show that the unit vector normal to the plane

$$ax + by + cz = d$$

is given by

$$\hat{\mathbf{n}} = \pm(\mathbf{i}a + \mathbf{j}b + \mathbf{k}c)/(a^2 + b^2 + c^2)^{1/2}.$$

(b) Explain in geometric terms why this expression for $\hat{\mathbf{n}}$ is independent of the constant d.

II–3 Derive expressions for the unit normal vector for surfaces given by $y = g(x, z)$ and by $x = h(y, z)$. Use each to rederive the expression for the normal to the plane given in Problem II–2.

II–4 In each of the following use Equation II-12 to evaluate the surface integral $\iint_s G(x, y, z)dS$.

(a) $G(x, y, z) = z$,
    where S is the portion of the plane $x + y + z = 1$ in the first octant.

(b) $G(x, y, z) = \dfrac{1}{1 + 4(x^2 + y^2)}$,
    where S is the portion of the paraboloid $z = x^2 + y^2$ between $z = 0$ and $z = 1$.

(c) $G(x, y, z) = (1 - x^2 - y^2)^{3/2}$,
    where S is the hemisphere $z = (1 - x^2 - y^2)^{1/2}$.

II–5 In each of the following use Equation II–13 to evaluate the surface integral $\iint_s \mathbf{F} \cdot \mathbf{n} \, dS$.

(a) $\mathbf{F}(x, y, z) = \mathbf{i}x - \mathbf{k}z$,

where S is the portion of the plane $x + y + 2z = 2$ in the first octant.

(b) $\mathbf{F}(x, y, z) = \mathbf{i}x + \mathbf{j}y + \mathbf{k}z$,

where S is the hemisphere $z = \sqrt{a^2 - x^2 - y^2}$.

(c) $\mathbf{F}(x, y, z) = \mathbf{j}y + \mathbf{k}$,

where S is the portion of the paraboloid $z = 1 - x^2 - y^2$ above the xy-plane.

II–6 The distribution of mass on the hemispherical shell

$$z = (R^2 - x^2 - y^2)^{1/2}$$

is given by

$$\sigma(x, y, z) = (\sigma_0/R^2)(x^2 + y^2).$$

where σ_0 is a constant. Find an expression in terms of σ_0 and R for the total mass of the shell.

II–7 Find the moment of inertia about the z-axis of the hemispherical shell of Problem II–6.

II–8 An electrostatic field is given by

$$\mathbf{E} = \lambda(\mathbf{i}yz + \mathbf{j}xz + \mathbf{k}xy),$$

where λ is a constant. Use Gauss' law to find the total charge enclosed by the surface shown in the figure consisting of S_1, the hemisphere

$$z = (R^2 - x^2 - y^2)^{1/2},$$

and S_2, its circular base in the xy-plane.

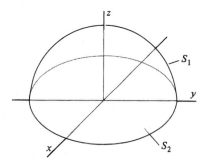

II–9 An electrostatic field is given by $\mathbf{E} = \lambda(\mathbf{i}x + \mathbf{j}y)$, where λ is a constant. Use Gauss' law to find the total charge enclosed by the surface shown in the figure consisting of S_1, the curved portion of the half-cylinder $z = (r^2 - y^2)^{1/2}$ of length h; S_2 and S_3, the two semicircular plane end pieces; and S_4, the rectangular portion of the xy-plane. Express your results in terms of λ, r, and h.

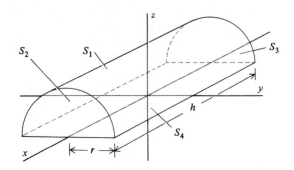

II–10 It sometimes happens that surface integrals can be evaluated without using the long-winded procedures outlined in the text. Try evaluating $\iint_S \mathbf{F} \cdot \hat{\mathbf{n}}\, dS$ for each of the following; think a bit and avoid a lot of work!

(a) $\mathbf{F} = \mathbf{i}x + \mathbf{j}y + \mathbf{k}z$.

S, the three squares each of side b as shown in the figure.

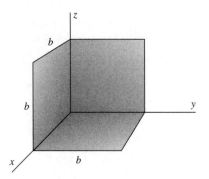

(b) $\mathbf{F} = (\mathbf{i}x + \mathbf{j}y) \ln (x^2 + y^2)$.

S, the cylinder (including the top and bottom) of radius R and height h shown in the figure.

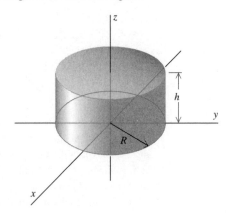

(c) $\mathbf{F} = (\mathbf{i}x + \mathbf{j}y + \mathbf{k}z)e^{-(x^2+y^2+z^2)}$.

    S, the surface of the sphere of radius R centered at the origin as shown in the figure.

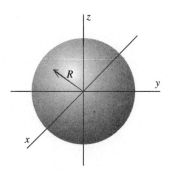

(d) $\mathbf{F} = \mathbf{i}E(x)$, where $E(x)$ is an arbitrary scalar function of x.

    S, the surface of the cube of side b shown in the figure.

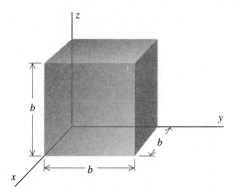

II–11 (a) Use Gauss' law and symmetry to find the electrostatic field as a function of position for an infinite uniform plane of charge. Let the charge lie in the yz-plane and denote the charge per unit area by σ.

(b) Repeat part (a) for an infinite slab of charge parallel to the yz-plane, whose density is given by

$$\rho(x) = \begin{cases} \rho_0, & -b < x < b, \\ 0, & |x| \geq b, \end{cases}$$

where ρ_0 and b are constants.

(c) Repeat part (b) with $\rho(x) = \rho_0 e^{-|x/b|}$.

II–12 (a) Use Gauss' law and symmetry to find the electrostatic field as a function of position for an infinite uniform line of charge. Let the charge lie along the z-axis and denote the charge per unit length by λ.

(b) Repeat part (a) for an infinite cylinder of charge whose axis

coincides with the z-axis and whose density is given in cylindrical coordinates by

$$\rho(r) = \begin{cases} \rho_0, & r < b, \\ 0, & r \ge b, \end{cases}$$

where ρ_0 and b are constants.

(c) Repeat part (b) with $\rho(r) = \rho_0 e^{-r/b}$.

II–13 (a) Use Gauss' law and symmetry to find the electrostatic field
as a function of position for the spherically symmetric charge
distribution whose density is given in spherical coordinates by

$$\rho(r) = \begin{cases} \rho_0, & r < b, \\ 0, & r \ge b, \end{cases}$$

where ρ_0 and b are constants.

(b) Repeat part (a) for $\rho(r) = \rho_0 e^{-r/b}$.

(c) Repeat part (a) for

$$\rho(r) = \begin{cases} \rho_0, & r < b, \\ \rho_1, & b \le r < 2b, \\ 0, & r \ge 2b. \end{cases}$$

How must ρ_0 and ρ_1 be related so that the field will be zero for
$r > 2b$? What is the total charge of this distribution under these
circumstances?

II–14 Calculate the divergence of each of the following functions using
Equation (II–22):

(a) $\mathbf{i}x^2 + \mathbf{j}y^2 + \mathbf{k}z^2$.
(b) $\mathbf{i}yz + \mathbf{j}xz + \mathbf{k}xy$.
(c) $\mathbf{i}e^{-x} + \mathbf{j}e^{-y} + \mathbf{k}e^{-z}$.
(d) $\mathbf{i} - 3\mathbf{j} + \mathbf{k}z^2$.
(e) $(-\mathbf{i}xy + \mathbf{j}x^2)/(x^2 + y^2)$, $(x, y) \ne (0, 0)$.
(f) $\mathbf{k}\sqrt{x^2 + y^2}$.
(g) $\mathbf{i}x + \mathbf{j}y + \mathbf{k}z$.
(h) $(-\mathbf{i}y + \mathbf{j}x)/\sqrt{x^2 + y^2}$, $(x, y) \ne (0, 0)$.

II–15 (a) Calculate $\iint_s \mathbf{F} \cdot \hat{\mathbf{n}} \, dS$ for the function in Problem II–14(a)
over the surface of a cube of side s whose center is at (x_0, y_0, z_0)
and whose faces are parallel to the coordinate planes.

(b) Divide the above result by the volume of the cube and calculate the limit of the quotient as $s \rightarrow 0$. Compare your result
with the divergence found in Problem II–14(a).

(c) Repeat parts (a) and (b) for the function of Problem II–14(b)
and (c).

Problems

II–16 (a) Calculate the divergence of the function

$$\mathbf{F}(x, y, z) = \mathbf{i}f(x) + \mathbf{j}f(y) + \mathbf{k}f(-2z)$$

and show that it is zero at the point $(c, c, -c/2)$.

(b) Calculate the divergence of

$$\mathbf{G}(x, y, z) = \mathbf{i}f(y, z) + \mathbf{j}g(x, z) + \mathbf{k}h(x, y).$$

II–17 In the text we obtained the result

$$\nabla \cdot \mathbf{F} = \frac{\partial F_x}{\partial x} + \frac{\partial F_y}{\partial y} + \frac{\partial F_z}{\partial z}$$

by integrating over the surface of a small rectangular parallelepiped. As an example of the fact that this result is independent of the surface, rederive it using the wedge-shaped surface shown in the figure.

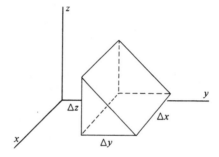

II–18 (a) Let $\mathbf{i}$, $\mathbf{j}$, and $\mathbf{k}$ be unit vectors in Cartesian coordinates and $\hat{\mathbf{e}}_r$, $\hat{\mathbf{e}}_\theta$, and $\hat{\mathbf{e}}_z$ be unit vectors in cylindrical coordinates. Show that

$$\mathbf{i} = \hat{\mathbf{e}}_r \cos \theta - \hat{\mathbf{e}}_\theta \sin \theta,$$

$$\mathbf{j} = \hat{\mathbf{e}}_r \sin \theta + \hat{\mathbf{e}}_\theta \cos \theta,$$

$$\mathbf{k} = \hat{\mathbf{e}}_z.$$

(b) Rewrite the function in Problem II–14(e) in cylindrical coordinates and compute its divergence, using Equation (II–24). Convert your result back to Cartesian coordinates and compare with the answer obtained in Problem II–14(e).

(c) Repeat part (b) for the function of Problem II–14(f).

II–19 (a) Let $\mathbf{i}$, $\mathbf{j}$, and $\mathbf{k}$ be unit vectors in Cartesian coordinates and $\hat{\mathbf{e}}_r$, $\hat{\mathbf{e}}_\theta$, and $\hat{\mathbf{e}}_\phi$ be unit vectors in spherical coordinates. Show that

$$\mathbf{i} = \hat{\mathbf{e}}_r \sin \theta \cos \phi + \hat{\mathbf{e}}_\theta \cos \theta \cos \phi - \hat{\mathbf{e}}_\phi \sin \phi,$$

$$\mathbf{j} = \hat{\mathbf{e}}_r \sin \theta \sin \phi + \hat{\mathbf{e}}_\theta \cos \theta \sin \phi + \hat{\mathbf{e}}_\phi \cos \phi,$$

$$\mathbf{k} = \hat{\mathbf{e}}_r \cos \theta - \hat{\mathbf{e}}_\theta \sin \theta.$$

[*Hint:* It's easier to express $\hat{e}_r$, $\hat{e}_\theta$, and $\hat{e}_\phi$ in terms of **i**, **j**, and **k** and then solve algebraically for **i**, **j**, and **k**. To do this, first use the fact that $\hat{e}_r = \mathbf{r}/r = (\mathbf{i}x + \mathbf{j}y + \mathbf{k}z)/r$. Next, reasoning geometrically, show that $\hat{e}_\phi = -\mathbf{i} \sin \phi + \mathbf{j} \cos \phi$. Finally, calculate $\hat{e}_\theta = \hat{e}_\phi \times \hat{e}_r$.]

(b) Rewrite the function of Problem II–14(g) in spherical coordinates and compute its divergence using Equation (II–25). Convert your result back to Cartesian coordinates and compare with the answer obtained in Problem II–14(g).

(c) Repeat part (b) for the function of Problem II–14(h).

II–20 In cylindrical coordinates the divergence of **F** is given by

$$\nabla \cdot \mathbf{F} = \frac{1}{r} \frac{\partial}{\partial r} (rF_r) + \frac{1}{r} \frac{\partial F_\theta}{\partial \theta} + \frac{\partial F_z}{\partial z}.$$

In the text (pages 41–42) we derived the first term of this expression. Proceeding the same way, obtain the other two terms.

II–21 Repeat Problem II–20 to obtain the divergence in spherical coordinates by carrying out the surface integral over the surface of the volume shown in the figure and thereby obtaining the expression

$$\nabla \cdot \mathbf{F} = \frac{1}{r^2} \frac{\partial}{\partial r} (r^2 F_r) + \frac{1}{r \sin \theta} \frac{\partial}{\partial \theta} (\sin \theta \, F_\theta) + \frac{1}{r \sin \theta} \frac{\partial F_\phi}{\partial \phi}.$$

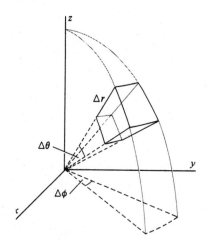

II–22 Consider a vector function of the form

$$\mathbf{F}(r) = \hat{e}_r f(r),$$

where $\hat{e}_r = (\mathbf{i}x + \mathbf{j}y + \mathbf{k}z)/r$ is the unit vector in the radial direction, $r = (x^2 + y^2 + z^2)^{1/2}$, and $f(r)$ is a differentiable scalar function. Using the results of Problem II–21, determine $f(r)$ so that $\nabla \cdot \mathbf{F} = 0$. A vector function whose divergence is zero is said to be *solenoidal*.

II–23 Verify the divergence theorem

$$\iint_S \mathbf{F} \cdot \hat{\mathbf{n}} \, dS = \iiint_V \nabla \cdot \mathbf{F} \, dV$$

in each of the following cases:

(a) $\mathbf{F} = \mathbf{i}x + \mathbf{j}y + \mathbf{k}z$.

    S, the surface of the cube of side b shown in the figure.

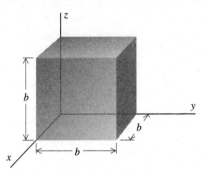

(b) $\mathbf{F} = \hat{\mathbf{e}}_r r + \hat{\mathbf{e}}_z z$,

    $\mathbf{r} = \mathbf{i}x + \mathbf{j}y$.

    S, the surface of the quarter cylinder (radius R, height h) shown in the figure.

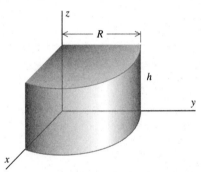

(c) $\mathbf{F} = \hat{\mathbf{e}}_r r^2$,

    $\mathbf{r} = \mathbf{i}x + \mathbf{j}y + \mathbf{k}z$.

    S, the surface of the sphere of radius R centered at the origin as shown in the figure.

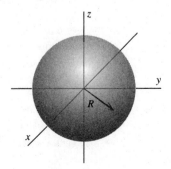

II–24 (a) One of Maxwell's equations states that $\nabla \cdot \mathbf{B} = 0$, where $\mathbf{B}$ is any magnetic field. Show that

$$\iint_S \hat{\mathbf{n}} \cdot \mathbf{B} \, dS = 0$$

for any closed surface S.

(b) Determine the flux of a uniform magnetic field $\mathbf{B}$ through the curved surface of a right circular cone (radius R, height h) oriented so that $\mathbf{B}$ is normal to the base of the cone as shown in the figure. (A uniform field is one which has the same magnitude and direction everywhere.)

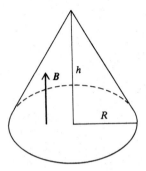

II–25 Use the divergence theorem to show that

$$\iint_S \hat{\mathbf{n}} \, dS = 0,$$

where S is a closed surface and $\hat{\mathbf{n}}$ the unit vector normal to the surface S.

II–26 (a) Use the divergence theorem to show that

$$\frac{1}{3} \iint_S \hat{\mathbf{n}} \cdot \mathbf{r} \, dS = V,$$

where S is a closed surface enclosing a region of volume V, $\hat{\mathbf{n}}$ is a unit vector normal to the surface S, and $\mathbf{r} = \mathbf{i}x + \mathbf{j}y + \mathbf{k}z$.

(b) Use the expression given in (a) to find the volume of:

(i) a rectangular parallelepiped with sides a, b, c.

(ii) a right circular cone with height h and base radius R.
     [*Hint:* The calculation is very simple with the cone oriented as shown in the figure]

(iii) a sphere of radius R.

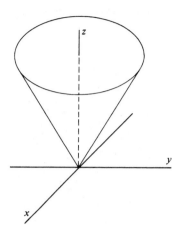

II–27 (a) Consider a vector function with the property $\nabla \cdot \mathbf{F} = 0$ everywhere on two closed surfaces S_1 and S_2 and in the volume V enclosed by them (see the figure). Show that the flux of $\mathbf{F}$ through S_1 equals the flux of $\mathbf{F}$ through S_2. In calculating the fluxes, choose the direction of the normals as indicated by the arrows in the figure.

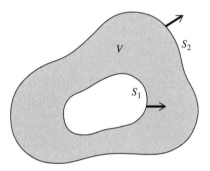

(b) Given the electrostatic field of a point charge q situated at $r = 0$,

$$\mathbf{E} = \frac{1}{4\pi\epsilon_0} \frac{q}{r^2} \hat{\mathbf{e}}_r,$$

where $r^2 = x^2 + y^2 + z^2$, show by direct calculation that

$$\nabla \cdot \mathbf{E} = 0, \quad \text{for all } r \neq 0.$$

(c) Prove Gauss' law for the field of a single point charge given in (b). [*Hint:* It is easy to calculate the flux of $\mathbf{E}$ over a sphere centered at $r = 0$.]

(d) How would you extend this proof to cover the case of an arbitrary charge distribution?

II–28 (a) Show by direct calculation that the divergence theorem does not hold for

$$\mathbf{F}(r, \theta, \phi) = \frac{\hat{\mathbf{e}}_r}{r^2},$$

with S the surface of a sphere of radius R centered at the origin, and V the enclosed volume. Why does the theorem fail?

(b) Verify by direct calculation that the divergence theorem does hold for the function $\mathbf{F}$ of part (a) when S is the surface S_1 of a sphere of radius R_1 plus the surface S_2 of a sphere of radius R_2, both centered at the origin, and V is the volume enclosed by S_1 and S_2.

(c) In general, what restriction must be placed on a surface S so that the divergence theorem will hold for the function of part (a)?

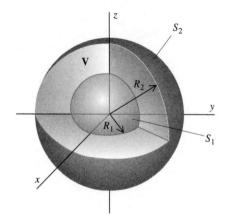

Chapter III

Line Integrals and the Curl

To err from the right path is common to Mankind.

Sophocles

Work and Line Integrals

We remarked above that the differential form of Gauss' law, Equations (II–18) and (II–23), although it fulfills our goal of relating a property of the electric field (its divergence) at a point to a known quantity (the charge density) at the same point, nonetheless falls short of providing a convenient way to find **E**. The reason is that $\nabla \cdot \mathbf{E} = \rho/\epsilon_0$ is (or seems to be) a single differential equation in three unknowns (E_x, E_y, E_z). But there is another feature of electrostatic fields which has not yet played an explicit role in our discussion and which will yield a relationship among the components of **E**. It will thus provide us with the crucial last step in obtaining a useful way to calculate fields. In the process

63

of examining this question, we shall encounter some of the most important topics in vector calculus.

The property of electrostatic fields that we shall now begin to discuss is intimately bound up with the question of work and energy. You no doubt recall the elementary definition of work as force times distance. Thus, in one dimension, if a force $F(x)$ acts from $x = a$ to $x = b$, the work done is, by definition,

$$\int_a^b F(x) \, dx.$$

To be able to handle more general situations, we must now introduce the concept of the *line integral.*

Suppose we have a curve C in three dimensions (Figure III–1) and suppose the curve is *directed.* By this we mean that we put

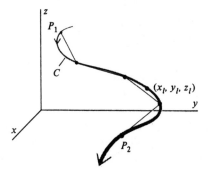

Figure III–1

an arrow on the curve and say "This is the positive direction." Let s be the arc length measured along the curve from some arbitrary point on it with $s = s_1$ at a point P_1 and $s = s_2$ at P_2. Suppose further that we have a function $f(x, y, z)$ defined everywhere on C. Now let us subdivide the portion of C between P_1 and P_2 arbitrarily into N sections. Figure III–1 shows an example of such a subdivision for $N = 4$. Next, join successive subdivision points by chords, a typical one of which, say the lth, has length Δs_l. Now evaluate $f(x, y, z)$ at (x_l, y_l, z_l), which is any point on the lth subdivision of the curve, and form the product $f(x_l, y_l, z_l) \, \Delta s_l$. Doing this for each of the N segments of C, we form the sum

$$\sum_{l=1}^{N} f(x_l, y_l, z_l) \, \Delta s_l.$$

By definition, the line integral of $f(x, y, z)$ along the curve C is the limit of this sum as the number of subdivisions N approaches infinity and the length of each chord approaches zero:

$$\int_C f(x, y, z)\, ds = \lim_{\substack{N \to \infty \\ \text{each } \Delta s_l \to 0}} \sum_{l=1}^{N} f(x_l, y_l, z_l)\, \Delta s_l.$$

To evaluate the line integral, we need to know the path C. Usually the most convenient way to specify this path is parametrically in terms of the arc length parameter s. Thus, we write $x = x(s)$, $y = y(s)$, and $z = z(s)$. In such a situation the line integral can be reduced to an ordinary definite integral:

$$\int_C f(x, y, z)\, ds = \int_{s_1}^{s_2} f[x(s), y(s), z(s)]\, ds.$$

An example of a line integral will be helpful here. For simplicity let us work in two dimensions and evaluate

$$\int_C (x + y)\, ds,$$

where C is the straight line from the origin to the point whose coordinates are $(1, 1)$ (Figure III–2). If (x, y) are the coordinates

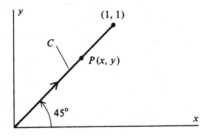

Figure III–2

of any point P on C and if s is the arc length measured from the origin, then $x = s/\sqrt{2}$ and $y = s/\sqrt{2}$. Hence, $x + y = 2s/\sqrt{2} = \sqrt{2}s$. Thus,

$$\int_C (x + y)\, ds = \sqrt{2} \int_0^{\sqrt{2}} s\, ds = \sqrt{2}.$$

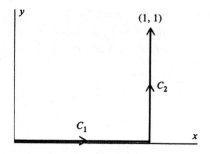

Figure III–3

Let us integrate this same function $(x + y)$ from $(0, 0)$ to $(1, 1)$ along another path as shown in Figure III–3. Here we break the integration into two parts, one along C_1 and the second along C_2. On C_1 we have $x = s$ and $y = 0$. Thus, on C_1, $x + y = s$, and so

$$\int_{C_1} (x + y)\, ds = \int_0^1 s\, ds = \tfrac{1}{2}.$$

Along C_2, $x = 1$ and $y = s$ [note that the arc length on this segment of the path is measured from the point $(1, 0)$]. It follows then that

$$\int_{C_2} (x + y)\, ds = \int_0^1 (1 + s)\, ds = \tfrac{3}{2}.$$

Adding the results for the two segments, we find

$$\int_C (x + y)\, ds$$

$$= \int_{C_1} (x + y)\, ds + \int_{C_2} (x + y)\, ds = \tfrac{1}{2} + \tfrac{3}{2} = 2.$$

The lesson to be learned is this: the value of a line integral can (indeed, usually does) depend on the path of integration.

Line Integrals Involving Vector Functions

Although the above discussion tells us what a line integral is, the kind of line integral we must deal with here has a feature not yet mentioned. You will recall that we introduced our discussion of line integrals with the concept of work. Work, in the most elementary sense, is force times displacement. That this needs elaboration becomes clear when we recognize that both force and displacement are vectors.

Thus, consider some path C in three dimensions (Figure III–4). Let us suppose that under the action of a force an object moves

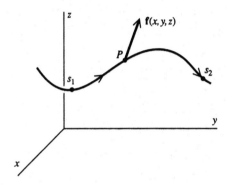

Figure III–4

on this path from s_1 to s_2. At any point P on the curve let the force acting be designated $\mathbf{f}(x, y, z)$. The component of $\mathbf{f}$ which does work is, by definition, only that one which acts *along* the curve, that is, the tangential component. Let $\hat{\mathbf{t}}$ denote a unit vector which is tangent to the curve at P.[1] Then the work done by the force in moving the object from s_1 to s_2 along the curve C is

$$W = \int_C \mathbf{f}(x,\ y,\ z) \cdot \hat{\mathbf{t}}\ ds,$$

where it is understood, of course, that the integration begins at $s = s_1$ and ends at $s = s_2$. The new feature of this integral is that the integrand is the dot product of two vector functions. To be able to handle such a line integral, we must know how to find $\hat{\mathbf{t}}$, and it is to this problem that we now turn.

Consider an arbitrary curve C (Figure III–5) parametrized by its arc length. At some point s on the curve we have $x = x(s)$, $y = y(s)$, and $z = z(s)$. At another point $s + \Delta s$ we have $x + \Delta x$

---

[1] $\hat{\mathbf{t}}$ is a function of x, y, and z and should really be written $\hat{\mathbf{t}}(x, y, z)$. We write simply $\hat{\mathbf{t}}$ to avoid complicating the notation.

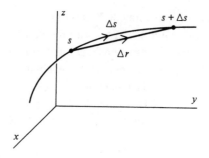

Figure III–5

$= x(s + \Delta s)$, $y + \Delta y = y(s + \Delta s)$, and $z + \Delta z = z(s + \Delta s)$. Thus, the chord joining the two points on the curve directed from the first to the second is the vector $\Delta \mathbf{r} \equiv \mathbf{i}\Delta x + \mathbf{j}\Delta y + \mathbf{k}\Delta z$, where

$$\Delta x = x(s + \Delta s) - x(s),$$
$$\Delta y = y(s + \Delta s) - y(s),$$
$$\Delta z = z(s + \Delta s) - z(s).$$

If we now divide this vector by Δs, we get

$$\frac{\Delta \mathbf{r}}{\Delta s} = \mathbf{i}\frac{\Delta x}{\Delta s} + \mathbf{j}\frac{\Delta y}{\Delta s} + \mathbf{k}\frac{\Delta z}{\Delta s}.$$

Taking the limit of this as Δs approaches zero yields

$$\mathbf{i}\frac{dx}{ds} + \mathbf{j}\frac{dy}{ds} + \mathbf{k}\frac{dz}{ds},$$

and we assert that this is $\hat{\mathbf{t}}$. To begin with, it's clear that as $\Delta s \to 0$, the vector $\Delta \mathbf{r}$ becomes tangent to the curve at s. Further, in the limit $\Delta s \to 0$, we see that $|\Delta \mathbf{r}| \to \Delta s$. Hence, in the limit the magnitude of this quantity is 1. It follows then that we can make the identification

$$\hat{\mathbf{t}}(s) = \mathbf{i}\frac{dx}{ds} + \mathbf{j}\frac{dy}{ds} + \mathbf{k}\frac{dz}{ds}.$$

If we return now to the expression for work W and use this formula for $\hat{\mathbf{t}}$, we find

$$W = \int_C \mathbf{f}(x, y, z) \cdot \left[\mathbf{i}\, \frac{dx}{ds} + \mathbf{j}\, \frac{dy}{ds} + \mathbf{k}\, \frac{dz}{ds} \right] ds$$

$$= \int_C (f_x\, dx + f_y\, dy + f_z\, dz).$$

This is a formal expression; often, to carry out the integration, it is useful to restore the ds as the following example illustrates. Consider

$$\mathbf{f}(x, y, z) = \mathbf{i}y - \mathbf{j}x$$

and the path shown in Figure III–6(a). To calculate $\int_C (\mathbf{f} \cdot \hat{\mathbf{t}})\, ds$

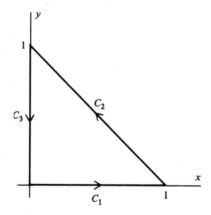

Figure III–6(a)

in this case, we break the path C into three parts, C_1, C_2, and C_3 as shown. Since $f_z = 0$, we have

$$\int_C (\mathbf{f} \cdot \hat{\mathbf{t}})\, ds = \int_C f_x\, dx + f_y\, dy$$

$$= \int_C y\, dx - x\, dy.$$

Now, on C_1, $y = 0$ and $dy = 0$, so there is no contribution to the integral. Similarly, on C_3 we have $x = 0$ and $dx = 0$, and again the result is zero. Thus, the only contribution to the integral over

C can come from C_2. Restoring the ds, we have

$$\int_C \left(y \frac{dx}{ds} - x \frac{dy}{ds} \right) ds.$$

But $(1 - x)/s = \cos 45° = 1/\sqrt{2}$ and $y/s = \sin 45° = 1/\sqrt{2}$ [Figure III–6(b)]. Thus,

$$\left. \begin{array}{l} x = 1 - \dfrac{s}{\sqrt{2}} \Rightarrow \dfrac{dx}{ds} = -\dfrac{1}{\sqrt{2}} \\[3mm] y = \dfrac{s}{\sqrt{2}} \Rightarrow \dfrac{dy}{ds} = \dfrac{1}{\sqrt{2}} \end{array} \right\} \quad 0 \le s \le \sqrt{2}.$$

Hence, the integral is

$$\int_0^{\sqrt{2}} \left[\frac{s}{\sqrt{2}} \left(-\frac{1}{\sqrt{2}} \right) - \left(1 - \frac{s}{\sqrt{2}} \right) \frac{1}{\sqrt{2}} \right] ds$$

$$= -\frac{1}{\sqrt{2}} \int_0^{\sqrt{2}} ds = -1.$$

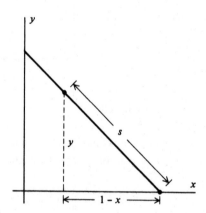

Figure III–6(b)

Path Independence

In a line integral the path of integration is one of the ingredients which determines the very function we integrate. It isn't remarkable, then, that the value of the integral can depend on the path of integration. What is remarkable is that, under some conditions, the value of the integral does *not* depend on the path!

We show how this path independence comes about in the case of the Coulomb force. Let a charge q_0 be fixed at the origin and let another charge q be situated at (x, y, z) (Figure III–7). The

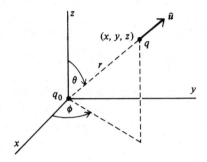

Figure III–7

Coulomb force on q is

$$\mathbf{F} = \frac{1}{4\pi\epsilon_0} \frac{qq_0}{r^2} \hat{\mathbf{u}}, \qquad \text{(III–1)}$$

where $r = (x^2 + y^2 + z^2)^{1/2}$ is the distance between the two charges and $\hat{\mathbf{u}}$ is a unit vector pointing from q_0 to q. With this arrangement $\hat{\mathbf{u}}$ is clearly in the radial direction. Even more clearly, the radial vector $\mathbf{r}$ is in the radial direction. Thus, we have $\hat{\mathbf{u}} = \mathbf{r}/r = (\mathbf{i}x + \mathbf{j}y + \mathbf{k}z)/r$, and so

$$\mathbf{F} = \frac{qq_0}{4\pi\epsilon_0} \frac{\mathbf{i}x + \mathbf{j}y + \mathbf{k}z}{r^3} .$$

Thus,

$$\mathbf{F} \cdot \hat{\mathbf{t}} \, ds = F_x \, dx + F_y \, dy + F_z \, dz$$

$$= \frac{qq_0}{4\pi\epsilon_0} \frac{x \, dx + y \, dy + z \, dz}{r^3}$$

The trick now is to use the relationship

$$r^2 = x^2 + y^2 + z^2.$$

Taking differentials in this equation and dividing by a factor of 2 yields

$$x \, dx + y \, dy + z \, dz = r \, dr,$$

so that

$$\mathbf{F} \cdot \hat{\mathbf{t}} \, ds = \frac{qq_0}{4\pi\epsilon_0} \frac{r \, dr}{r^3} = \frac{qq_0}{4\pi\epsilon_0} \frac{dr}{r^2}.$$

Suppose now that the charge q moves from a point P_1 at a distance r_1 from the origin to a point P_2 at a distance r_2, over some path C connecting the two points (Figure III–8). Then

$$\int_C \mathbf{F} \cdot \hat{\mathbf{t}} \, ds = \frac{qq_0}{4\pi\epsilon_0} \int_{r_1}^{r_2} \frac{dr}{r^2} = \frac{qq_0}{4\pi\epsilon_0} \left(\frac{1}{r_1} - \frac{1}{r_2} \right).$$

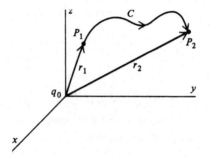

Figure III–8

Notice that to get this result, we haven't had to specify C in any way whatever; we'd get the same answer for *any* path connecting P_1 and P_2. This, of course, proves that the line integral

$$\int_C \mathbf{F} \cdot \hat{\mathbf{t}} \, ds$$

with $\mathbf{F}$ given by Equation (III–1) is path-independent, but the result, so far, has been established only for the Coulomb force on q due to a single charge q_0 [Equation (III–1)]. If there are many charges $q_1, q_2, \ldots, q_N$, then the total force on q is $\mathbf{F}_1 + \mathbf{F}_2 + \cdots + \mathbf{F}_N$ where $\mathbf{F}_l$ is the Coulomb force on q due to the lth charge q_i. Hence,

$$\int_C \mathbf{F} \cdot \hat{\mathbf{t}} \, ds = \int_C \mathbf{F}_1 \cdot \hat{\mathbf{t}} \, ds + \cdots + \int_C \mathbf{F}_N \cdot \hat{\mathbf{t}} \, ds.$$

Now the discussion given above shows that each term of this sum is path-independent; hence, so is the sum itself. (All this, of course, is merely an application of the superposition principle.) To phrase this result in terms of the field requires one last trivial step: Since $\mathbf{F} = q\mathbf{E}$, it follows that $q \int_C \mathbf{E} \cdot \hat{\mathbf{t}} \, ds$ is path-independent, whence $\int_C \mathbf{E} \cdot \hat{\mathbf{t}} \, ds$ is also. Strange to say, it is this fact that will enable us eventually to convert $\nabla \cdot \mathbf{E} = \rho/\epsilon_0$ into a more useful equation.

If you examine the foregoing discussion carefully, you'll see that the fact that the Coulomb force varies inversely as the square of r has nothing whatever to do with the path independence of the line integral. The path independence rests solely on two properties of the Coulomb force: (1) It depends only on the distance between the two particles, and (2) it acts along the line joining them. Any force $\mathbf{F}$ with these two properties is called a *central force*, and $\int_C \mathbf{F} \cdot \hat{\mathbf{t}} \, ds$ is independent of path for *any* central force.[2]

One further step pertaining to path independence can be taken here. If

$$\int_C \mathbf{F} \cdot \hat{\mathbf{t}} \, ds$$

is independent of path, then

$$\int_{C_1} \mathbf{F} \cdot \hat{\mathbf{t}} \, ds = \int_{C_2} \mathbf{F} \cdot \hat{\mathbf{t}} \, ds,$$

where, as indicated in Figure III–9, C_1 and C_2 are two different arbitrary paths connecting the two points P_1 and P_2 and directed as shown in the figure. Now if instead of integrating along C_1 from P_1 to P_2, we go the other way, we simply change the sign

---

[2] Our having illustrated path independence with a central force may give the erroneous impression that *only* central forces have path-independent line integrals. That is certainly *not* true; many functions which are not central forces have path-independent line integrals. Later we'll develop a simple criterion for identifying such functions.

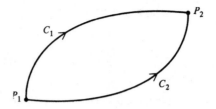

Figure III–9

of the line integral; that is,

$$\int_{-C_1} \mathbf{F} \cdot \hat{\mathbf{t}} \, ds = - \int_{C_1} \mathbf{F} \cdot \hat{\mathbf{t}} \, ds,$$

where ''$-C_1$'' merely indicates that the integration is to be carried out along C_1 from P_2 to P_1. Thus

$$\int_{C_2} \mathbf{F} \cdot \hat{\mathbf{t}} \, ds = - \int_{-C_1} \mathbf{F} \cdot \hat{\mathbf{t}} \, ds$$

or

$$\int_{-C_1 + C_2} \mathbf{F} \cdot \hat{\mathbf{t}} \, ds = 0.$$

But ''$-C_1 + C_2$'' is just the closed loop from P_1 to P_2 and back, as shown in Figure III–10. Thus, if $\int \mathbf{F} \cdot \hat{\mathbf{t}} \, ds$ is independent of

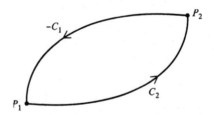

Figure III–10

path, then

$$\oint \mathbf{F} \cdot \hat{\mathbf{t}} \, ds = 0,$$

where $\oint$ is the standard notation for a line integral around a closed path. It follows that if **E** is an electrostatic field, we can write

$$\oint \mathbf{E} \cdot \hat{\mathbf{t}} \, ds = 0. \qquad \text{(III–2)}$$

circulation

The term "circulation" is often given to the path integral around a closed curve of the tangential component of a vector function. Thus we have demonstrated that the circulation of the electrostatic field is zero. In what follows we'll call this the circulation law.

The Curl

If we are given some vector function $\mathbf{F}(x, y, z)$ and asked "Could this be an electrostatic field?" we can, in principle, provide an answer. If

$$\oint \mathbf{F} \cdot \hat{\mathbf{t}} \, ds \neq 0$$

over even *one* path, then **F** cannot be an electrostatic field. If

$$\oint \mathbf{F} \cdot \hat{\mathbf{t}} \, ds = 0$$

over *every* closed path, then **F** can (but does not have to) be an electrostatic field.

Clearly this criterion is not easy to apply since we must be sure the circulation of **F** is zero over all possible paths. To develop a more useful criterion, we proceed much as we did in dealing with Gauss' law, which, like the circulation law, is an expression involving an integral over the electric field. Gauss' law is more useful in the differential form [Equations (II–18) and (II–23)] obtained by considering the ratio of flux to volume for ever decreasing surfaces. We now treat the circulation law in the same spirit and attempt to find the differential form of Equation (III–2). To stress the generality of our analysis and results, we deal with an arbitrary function $\mathbf{F}(x, y, z)$ and specialize to $\mathbf{E}(x, y, z)$ at a later stage in the development.

Let us consider the circulation of **F** over a small rectangle parallel to the *xy*-plane, with sides Δx and Δy and with the point

(x, y, z) at the center [Figure III–11(a)]. As shown in Figure III–11(b), we carry out the path integration in a counterclockwise direction looking down at the xy-plane. The line integral is broken up into four parts: C_B (bottom), C_R (right), C_T (top), and C_L

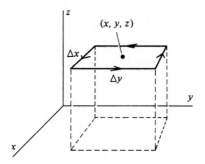

Figure III–11(a)

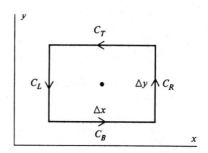

Figure III–11(b)

(left). Since the rectangle is small (eventually we shall take the limit as it shrinks down to zero), we'll approximate the integral over each segment by $\mathbf{F} \cdot \hat{\mathbf{t}}$ evaluated at the center of the segment, multiplied by the length of the segment.[3]

Taking C_B first, we have

$$\int_{C_B} \mathbf{F} \cdot \hat{\mathbf{t}} \, ds = \int_{C_B} F_x \, dx$$

$$\simeq F_x \left(x, y - \frac{\Delta y}{2}, z \right) \Delta x. \quad \text{(III–3a)}$$

[3] Reread footnote 9 of Chapter II and then give an argument in support of this approximation.

Over C_T we find

$$\int_{C_T} \mathbf{F} \cdot \hat{\mathbf{t}} \, ds = \int_{C_T} F_x \, dx$$

$$\simeq -F_x \left(x, y + \frac{\Delta y}{2}, z \right) \Delta x. \quad \text{(III–3b)}$$

The negative sign here is required by the fact that

$$\int_{C_T} F_x \, dx = \int_{C_T} F_x \frac{dx}{ds} \, ds$$

and $dx/ds = -1$ over C_T. Adding Equation (III–3a) and (III–3b), we find

$$\int_{C_T + C_B} (\mathbf{F} \cdot \hat{\mathbf{t}}) \, ds$$

$$\simeq - \left[F_x \left(x, y + \frac{\Delta y}{2}, z \right) - F_x \left(x, y - \frac{\Delta y}{2}, z \right) \right] \Delta x$$

$$\simeq - \frac{F_x \left(x, y + \frac{\Delta y}{2}, z \right) - F_x \left(x, y - \frac{\Delta y}{2}, z \right)}{\Delta y} \Delta x \, \Delta y.$$

The factor $\Delta x \, \Delta y$ is clearly the area ΔS of the rectangle. Thus,

$$\frac{1}{\Delta S} \int_{C_T + C_B} (\mathbf{F} \cdot \hat{\mathbf{t}}) \, ds \qquad \qquad \text{(III–4)}$$

$$\simeq - \frac{F_x \left(x, y + \frac{\Delta y}{2}, z \right) - F_x \left(x, y - \frac{\Delta y}{2}, z \right)}{\Delta y}.$$

Exactly the same sort of analysis applied to the left and right sides of the rectangle (C_L and C_R) results in

$$\frac{1}{\Delta S} \int_{C_L + C_R} (\mathbf{F} \cdot \hat{\mathbf{t}}) \, ds$$

$$\simeq \frac{F_y \left(x + \frac{\Delta x}{2}, y, z \right) - F_y \left(x - \frac{\Delta x}{2}, y, z \right)}{\Delta x}. \quad \text{(III–5)}$$

Adding Equations (III–4) and (III–5) and taking the limit as ΔS shrinks down about a point (x, y, z) (in which case Δx and Δy $\to 0$ as well), we get

$$\lim_{\substack{\Delta S \to 0 \\ \text{about } (x,y,z)}} \frac{1}{\Delta S} \oint \mathbf{F} \cdot \hat{\mathbf{t}}\, ds = \frac{\partial F_y}{\partial x} - \frac{\partial F_x}{\partial y}, \qquad \text{(III–6)}$$

where $\oint$ is our semicomical notation meaning the circulation around the little rectangle.

You may wonder about the generality and uniqueness of this result since it is obtained using a path of integration which is special in two ways: first, it is a rectangle, and second, it is parallel to the xy-plane. If the path were not a rectangle, but a plane curve of arbitrary shape, it would not affect our result (see Problems III–2 and III–30). But our result definitely *does* depend on the special orientation of the path of integration. The choice of orientation made above clearly suggests two others, and they are shown in Figure III–12(a) and (b) along with the result of

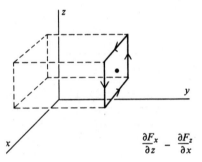

$$\frac{\partial F_x}{\partial z} - \frac{\partial F_z}{\partial x}$$

Figure III–12(a)

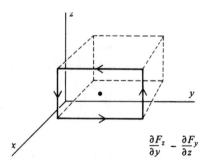

$$\frac{\partial F_z}{\partial y} - \frac{\partial F_y}{\partial z}$$

Figure III–12(b)

calculating

$$\lim_{\substack{\Delta S \to 0 \\ \text{about } (x,y,z)}} \frac{1}{\Delta S} \oint \mathbf{F} \cdot \hat{\mathbf{t}} \, ds$$

for each.

Each of these three paths is named in honor of the vector normal to the enclosed area. The convention we use is this: Trace the curve C so that the enclosed area is always to the left [Figure III–13(a)]. Then choose the normal so that it points ''up'' in the

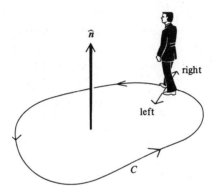

Figure III–13(a)

direction shown in the picture. This convention is sometimes called the right-hand rule, for if the right hand is oriented so that the fingers curl in the direction in which the curve is traced, the thumb, extended, points in the direction of the normal [Figure III–13(b)]. Using the right-hand rule, we have the following:

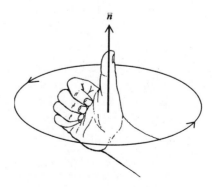

Figure III–13(b)

for a path whose normal is **i**, we get $\dfrac{\partial F_z}{\partial y} - \dfrac{\partial F_y}{\partial z}$,

for a path whose normal is **j**, we get $\dfrac{\partial F_x}{\partial z} - \dfrac{\partial F_z}{\partial x}$,   (III–7a)

for a path whose normal is **k**, we get $\dfrac{\partial F_y}{\partial x} - \dfrac{\partial F_x}{\partial y}$.

It turns out that these three quantities are the Cartesian components of a vector. To this vector we give the name "curl of **F**," which we write curl **F**. Thus, we have

$$\text{curl } \mathbf{F} = \mathbf{i} \left(\frac{\partial F_z}{\partial y} - \frac{\partial F_y}{\partial z} \right) + \mathbf{j} \left(\frac{\partial F_x}{\partial z} - \frac{\partial F_z}{\partial x} \right)$$

$$+ \mathbf{k} \left(\frac{\partial F_y}{\partial x} - \frac{\partial F_x}{\partial y} \right) . \quad \text{(III–7b)}$$

This expression is often (indeed, usually) given as the definition of the curl, but we prefer to regard it as merely the form of the curl in Cartesian coordinates. We shall define the curl as the limit of circulation to area as the area tends to zero. To be precise, let $\oint_{C_n} \mathbf{F} \cdot \hat{\mathbf{t}} \, ds$ be the circulation of **F** about some path whose normal is $\hat{\mathbf{n}}$ as shown in Figure III–14. Then by definition

$$\hat{\mathbf{n}} \cdot \text{curl } \mathbf{F} = \lim_{\substack{\Delta S \to 0 \\ \text{about } (x,y,z)}} \frac{1}{\Delta S} \oint_{C_n} \mathbf{F} \cdot \hat{\mathbf{t}} \, ds. \quad \text{(III–8)}$$

By taking $\hat{\mathbf{n}}$ successively equal to **i**, **j**, and **k**, we get back the results given in Equation (III–7b). Since this limit will, in general, have different values for different points (x, y, z), the curl of **F** is a *vector function* of position.[4] Note incidentally that although in our work we always assumed that the area enclosed

[4] The word *rotation* (abbreviated "rot," amusingly enough) was once used for what we now call the *curl*. Though the term has long since dropped out of use, a related one survives: If curl **F** = 0, the function **F** is said to be *irrotational*.

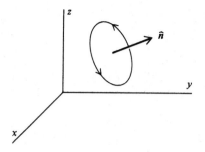

Figure III–14

by the path of integration was a plane, this need not be the case. Since the curl is defined in terms of a limit in which the enclosed surface shrinks to zero about some point, in the final stages of this limiting process the enclosed surface is infinitesimally close to a plane, and all our considerations apply.

Since it is undoubtedly beyond the powers of a mere mortal to remember the expression given above for curl **F** in Cartesian coordinates [Equation (III–7b)], it is fortunate that there is a mnemonic device to fall back on. If the three-by-three determinant

$$\begin{vmatrix} \mathbf{i} & \mathbf{j} & \mathbf{k} \\ \partial/\partial x & \partial/\partial y & \partial/\partial z \\ F_x & F_y & F_z \end{vmatrix}$$

is expanded (most conveniently in minors of the first row) and if certain "products" are interpreted as partial derivatives [for example, $(\partial/\partial x)F_y = \partial F_y/\partial x$], the result will be identical with the one given in Equation (III–7b).[5] Thus, the anguish of remembering the form of curl **F** in Cartesian coordinates can be replaced by the pain of remembering how to expand a three-by-three determinant. *Chacun à son goût.*

As an example of calculating the curl, consider the vector function

$$\mathbf{F}(x, y, z) = \mathbf{i}xz + \mathbf{j}yz - \mathbf{k}y^2.$$

[5] A mathematician would object to this since, strictly speaking, a determinant cannot contain either vectors or operators. We aren't doing any serious damage, however, because our "determinant" is merely a memory aid.

We have

$$\text{curl } \mathbf{F} = \begin{vmatrix} \mathbf{i} & \mathbf{j} & \mathbf{k} \\ \partial/\partial x & \partial/\partial y & \partial/\partial z \\ xz & yz & -y^2 \end{vmatrix}$$

$$= \mathbf{i}(-2y - y) + \mathbf{j}(x - 0) + \mathbf{k}(0 - 0)$$

$$= -3\mathbf{i}y + \mathbf{j}x.$$

You may have noticed that the curl operator can be written in terms of the del notation we introduced earlier. You can verify for yourself that

$$\text{curl } \mathbf{F} = \nabla \times \mathbf{F},$$

which is read "del cross **F**." Henceforth, we shall always use $\nabla \times \mathbf{F}$ to indicate the curl.

The Curl in Cylindrical and Spherical Coordinates

To obtain the form of $\nabla \times \mathbf{F}$ in other coordinate systems, we proceed as we did above in finding the Cartesian form, merely modifying the paths of integration appropriately. As an example, using the path shown in Figure III–15(a) will yield the z-component of $\nabla \times \mathbf{F}$ in cylindrical coordinates.[6] Note that we trace

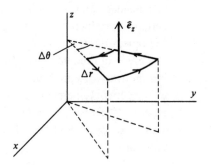

Figure III–15(a)

[6] In deriving the Cartesian form of $\nabla \times \mathbf{F}$, each segment of each path of integration (see Figures III–11 and III–12) was of the form x = constant, y = constant, or z = constant. Similarly, in deriving the cylindrical form, each segment of each path is of the form r = constant, θ = constant, or z = constant.

the curve in accordance with the right-hand rule given in the previous section. Viewing the path from above [as we do in

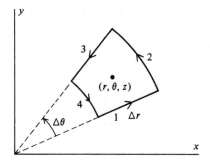

Figure III–15(b)

Figure III–15(b)], the line integral of $\mathbf{F}(r, \theta, z) \cdot \hat{\mathbf{t}}$ along the segment of path marked "1" is

$$\int_{C_1} \mathbf{F} \cdot \hat{\mathbf{t}}\ ds \simeq F_r \left(r, \theta - \frac{\Delta\theta}{2}, z \right) \Delta r,$$

while along segment "3" it is

$$\int_{C_3} \mathbf{F} \cdot \hat{\mathbf{t}}\ ds \simeq -F_r \left(r, \theta + \frac{\Delta\theta}{2}, z \right) \Delta r.$$

The area enclosed by the path is $r\ \Delta r\ \Delta\theta$, so

$$\frac{1}{\Delta S} \int_{C_1 + C_3} \mathbf{F} \cdot \hat{\mathbf{t}}\ ds$$

$$\simeq - \frac{\Delta r}{r\ \Delta r\ \Delta\theta} \left[F_r \left(r, \theta + \frac{\Delta\theta}{2}, z \right) - F_r \left(r, \theta - \frac{\Delta\theta}{2}, z \right) \right].$$

In the limit as Δr and $\Delta\theta$ tend to zero, this becomes

$$- \frac{1}{r} \frac{\partial F_r}{\partial \theta}$$

evaluated at the point (r, θ, z).

   Along segment "2" we find

$$\int_{C_2} \mathbf{F} \cdot \hat{\mathbf{t}}\ ds \simeq F_\theta \left(r + \frac{\Delta r}{2}, \theta, z \right) \left(r + \frac{\Delta r}{2} \right) \Delta\theta,$$

and along segment "4"

$$\int_{C_4} \mathbf{F} \cdot \hat{\mathbf{t}}\ ds \simeq -F_\theta \left(r - \frac{\Delta r}{2}, \theta, z \right) \left(r - \frac{\Delta r}{2} \right) \Delta\theta.$$

Thus,

$$\frac{1}{\Delta S} \int_{C_2 + C_4} \mathbf{F} \cdot \hat{\mathbf{t}}\ ds$$

$$\simeq \frac{\Delta\theta}{r\,\Delta r\,\Delta\theta} \left[\left(r + \frac{\Delta r}{2} \right) F_\theta \left(r + \frac{\Delta r}{2}, \theta, z \right) \right.$$

$$\left. - \left(r - \frac{\Delta r}{2} \right) F_\theta \left(r - \frac{\Delta r}{2}, \theta, z \right) \right].$$

In the limit this becomes $(1/r)(\partial/\partial r)(rF_\theta)$ evaluated at (r, θ, z). Hence,

$$(\nabla \times \mathbf{F})_z \equiv \lim_{\Delta S \to 0} \frac{1}{\Delta S} \oint \mathbf{F} \cdot \hat{\mathbf{t}}\ ds = \frac{1}{r} \frac{\partial}{\partial r} (rF_\theta) - \frac{1}{r} \frac{\partial F_r}{\partial \theta}.$$

Paths for finding the r- and θ-components of $\nabla \times \mathbf{F}$ are shown in Figures III–15(c) and (d), respectively. You are asked to obtain

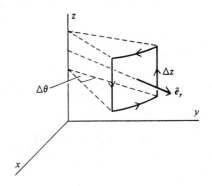

Figure III–15(c)

these two components yourself in Problem III–8. For completeness we give all three components of $\nabla \times \mathbf{F}$ in cylindrical

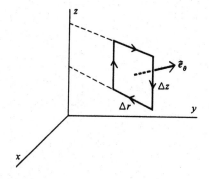

Figure III–15(d)

coordinates:

$$(\nabla \times \mathbf{F})_r = \frac{1}{r} \frac{\partial F_z}{\partial \theta} - \frac{\partial F_\theta}{\partial z},$$

$$(\nabla \times \mathbf{F})_\theta = \frac{\partial F_r}{\partial z} - \frac{\partial F_z}{\partial r},$$

$$(\nabla \times \mathbf{F})_z = \frac{1}{r} \frac{\partial}{\partial r} (rF_\theta) - \frac{1}{r} \frac{\partial F_r}{\partial \theta}.$$

The three components of curl $\mathbf{F}$ in *spherical* coordinates (see Problem III–9) are as follows:

$$(\nabla \times \mathbf{F})_r = \frac{1}{r \sin \theta} \frac{\partial}{\partial \theta} (\sin \theta \, F_\phi) - \frac{1}{r \sin \theta} \frac{\partial F_\theta}{\partial \phi},$$

$$(\nabla \times \mathbf{F})_\theta = \frac{1}{r \sin \theta} \frac{\partial F_r}{\partial \phi} - \frac{1}{r} \frac{\partial}{\partial r} (rF_\phi),$$

$$(\nabla \times \mathbf{F})_\phi = \frac{1}{r} \frac{\partial}{\partial r} (rF_\theta) - \frac{1}{r} \frac{\partial F_r}{\partial \theta}.$$

The Meaning of the Curl

The preceding discussion may leave you with the feeling that knowing how to define and calculate the curl of some vector function is a far cry from knowing what it is. The fact that the curl has something to do with a line integral *around* a closed path

85

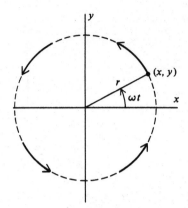

Figure III–16

(indeed, the word "curl" itself) may suggest to you that it somehow has to do with things rotating, swirling, or curling around. By means of a few examples taken from fluid motion, we'll try to make these vague impressions a little clearer.

Suppose water is flowing in circular paths, something like the water draining from a bathtub. A small volume of the water at a point (x, y) at time t has coordinates $x = r \cos \omega t$, $y = r \sin \omega t$, where ω is the constant angular velocity of the water (Figure III–16).[7] Thus, its velocity at (x, y) is

$$\mathbf{v} = \mathbf{i}(dx/dt) + \mathbf{j}(dy/dt) = r\omega[-\mathbf{i} \sin \omega t + \mathbf{j} \cos \omega t]$$
$$= \omega(-\mathbf{i}y + \mathbf{j}x).$$

This expression gives what is called the velocity field of the water; it tells us the velocity of the water at any point (x, y). Your intuition probably tells you that, because the motion is circular, this velocity must have a nonzero curl. In fact, as you can show very easily,

$$\nabla \times \mathbf{v} = 2\mathbf{k}\omega.$$

This result should seem quite reasonable because it says that curl of the velocity is proportional to the angular velocity of the swirling water. We see that $\nabla \times \mathbf{v}$ is a vector perpendicular to the

[7] This is not a realistic description of water draining from a tub since rotating water shears tangentially and its angular velocity will therefore vary with r. The crude description we use here is adequate for our purposes and has the virtue of being simple.

plane of motion and in the positive z-direction [Figure III–17(a)].

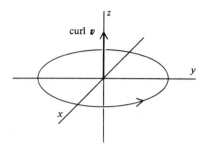

Figure III–17(a)

If the water were rotating around in the other direction, the curl of **v** would then be in the negative z-direction [Figure III–17(b)].

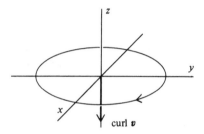

Figure III–17(b)

Note that this is consistent with the right-hand rule (see page 79). If we were to put a small paddle wheel in the water, it would commence spinning because the impinging water would exert a net torque on the paddles (Figure III–18). Furthermore, the pad-

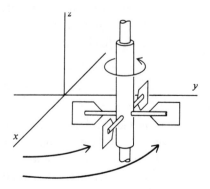

87   Figure III–18

dle wheel would rotate with its axis pointing in the direction of the curl.

Now consider a different velocity field, namely,

ex2

$$\mathbf{v} = \mathbf{j}v_0 e^{-y^2/\lambda^2},$$

where v_0 and λ are constants. Water with such a velocity field would have a flow pattern as indicated in Figure III–19. The velocity at all points is in the positive *y*-direction, and its magnitude (indicated by the length of the arrows) varies with *y*. Since you see only straight line flow here without any rotational motion, you would probably guess that $\nabla \times \mathbf{v} = 0$ in this case, and you would be right, as a simple calculation shows. There would be no net torque on a paddle wheel placed anywhere in this flow pattern, and as a consequence, it would not spin.[8]

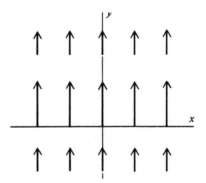

Figure III–19

ex3

Our last example is trickier than the two given above and shows that intuition can lead you astray if you're not careful. Let a velocity field be given by

$$\mathbf{v} = \mathbf{j}v_0 e^{-x^2/\lambda^2}.$$

As in the previous example, the velocity in this case is everywhere in the *y*-direction, but now it varies with *x*, not *y* (Figure III–20). Here as in the above example you see no evidence of rotational motion and you might guess that $\nabla \times \mathbf{v} = 0$ once again.

88

[8] If $\nabla \times \mathbf{v} = 0$, the flow is said to be irrotational. Compare with footnote 4.

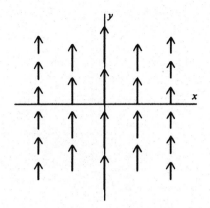

Figure III–20

But as you should show for yourself,

$$\nabla \times \mathbf{v} = -\mathbf{k}v_0 \frac{2x}{\lambda^2} e^{-x^2/\lambda^2}.$$

A small paddle wheel placed in this flow pattern would spin, even though the water is everywhere moving in the same direction. The reason this happens is that the velocity of the water varies with x, so that it strikes one of the paddles (P in Figure III–21)

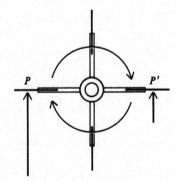

Figure III–21

with greater velocity than the other (P'). Thus, there will be a net torque. In more mathematical terms, the line integral of $\mathbf{v} \cdot \hat{\mathbf{t}}$ around a small rectangle (Figure III–22) will be different from zero, for while

$$\int_{\text{bottom}} \mathbf{v} \cdot \hat{\mathbf{t}} \, ds = \int_{\text{top}} \mathbf{v} \cdot \hat{\mathbf{t}} \, ds = 0,$$

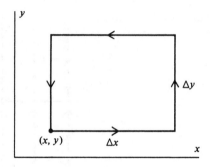

Figure III–22

the contributions from the other two sides are

$$\int_{\text{right}} \mathbf{v} \cdot \hat{\mathbf{t}} \, ds \simeq v_y(x + \Delta x) \, \Delta y$$

and

$$\int_{\text{left}} \mathbf{v} \cdot \hat{\mathbf{t}} \, ds \simeq -v_y(x) \, \Delta y.$$

These do not cancel because $v_y(x) \neq v_y(x + \Delta x)$. Incidentally, you should try to explain to your own satisfaction why in this example $\nabla \times \mathbf{v}$ is in the negative (positive) z-direction when x is positive (negative) and why $\nabla \times \mathbf{v} = 0$ at $x = 0$.

Differential Form of the Circulation Law

The curl is defined to be the limit of circulation to area. Thus,

$$\hat{\mathbf{n}} \cdot \nabla \times \mathbf{E} = \lim_{\Delta S \to 0} \frac{1}{\Delta S} \oint_C \mathbf{E} \cdot \hat{\mathbf{t}} \, ds,$$

where $\hat{\mathbf{n}}$ is a unit vector normal to the surface enclosed by C at the point about which the curve shrinks to zero. But if $\mathbf{E}$ is an electrostatic field, then

$$\oint_C \mathbf{E} \cdot \hat{\mathbf{t}} \, ds = 0$$

for any path C. It follows that

$$\hat{\mathbf{n}} \cdot \nabla \times \mathbf{E} = 0.$$

Since the curve C is arbitrary, we can arrange matters so that $\hat{\mathbf{n}}$ is a unit vector pointing in any direction we choose. Thus,

taking $\hat{\mathbf{n}} = \mathbf{i}$, we have $(\nabla \times \mathbf{E})_x = 0$;
taking $\hat{\mathbf{n}} = \mathbf{j}$, we have $(\nabla \times \mathbf{E})_y = 0$;
taking $\hat{\mathbf{n}} = \mathbf{k}$, we have $(\nabla \times \mathbf{E})_z = 0$.

Thus, all three Cartesian components of $\nabla \times \mathbf{E}$ vanish, and we can conclude that for an electrostatic field,

$$\nabla \times \mathbf{E} = 0.$$

This is the long-sought-after differential form of the circulation law. We are now in a position to give an alternative and much more tractable answer to the question ''Can a given vector function $\mathbf{F}(x, y, z)$ be an electrostatic field?'' The answer is:

If $\nabla \times \mathbf{F} = 0$, then $\mathbf{F}$ can be an electrostatic field, and
if $\nabla \times \mathbf{F} \neq 0$, then $\mathbf{F}$ cannot be an electrostatic field.

This is clearly a much more convenient criterion to apply than our earlier one (page 76), which required us to determine the line integral of $\mathbf{F}$ over *all* closed paths! To see how it works, let us do several examples.

Example 1. Could $\mathbf{F} = K(\mathbf{i}y + \mathbf{j}x)$ be an electrostatic field? (K is a constant.) Here we have

$$\frac{1}{K} \nabla \times \mathbf{F}$$

$$= \mathbf{i}\left(\frac{\partial}{\partial y} 0 - \frac{\partial x}{\partial z}\right) + \mathbf{j}\left(\frac{\partial y}{\partial z} - \frac{\partial}{\partial x} 0\right) + \mathbf{k}\left(\frac{\partial x}{\partial x} - \frac{\partial y}{\partial y}\right)$$

$$= 0 \Rightarrow \nabla \times \mathbf{F} = 0.$$

Answer: Yes.

Example 2. Could $\mathbf{F} = K(\mathbf{i}y - \mathbf{j}x)$ be an electrostatic field? In this case

$$\frac{1}{K} \nabla \times \mathbf{F} = \mathbf{k}\left(-\frac{\partial x}{\partial x} - \frac{\partial y}{\partial y}\right) = -2\mathbf{k} \Rightarrow \nabla \times \mathbf{F} = -2\mathbf{k}K.$$

Answer: No.

From these examples we can see how easy this criterion is to apply.

Stokes' Theorem

For the remainder of this chapter we digress from our presentation to discuss another famous theorem, one strongly reminiscent of the divergence theorem and yet, as we'll see, quite different from it. This theorem, named for the mathematician Stokes, relates a line integral around a closed path to a surface integral over what is called a *capping surface* of the path, so the first item on our agenda is to define this term. Suppose we have a closed curve C, as shown in Figure III–23(a), and imagine that it is made

Figure III–23(a)

of wire. Now let us suppose we attach an elastic membrane to the wire as indicated in Figure III–23(b). This membrane is a

Figure III–23(b)

capping surface of the curve C. Any other surface which can be formed by stretching the membrane is also a capping surface; an example is shown in Figure III–23(c). Figure III–24 shows four

Figure III–23(c)

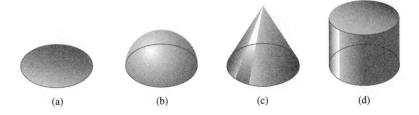

| (a) | (b) | (c) | (d) |

Figure III–24

different capping surfaces of a plane circular path: (a) the region of the plane enclosed by the circle, (b) a hemisphere with the circle as its rim, (c) the curved surface of a dunce cap (a right circular cone), and (d) the upper and lateral surfaces of a tuna fish can.

With these preliminary remarks in mind, you won't be surprised to see us begin this discussion of Stokes' theorem by considering some closed curve C and a capping surface S [Figure III–25(a)]. As we have done before, we approximate this capping

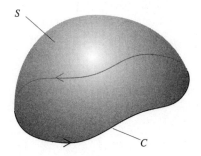

Figure III–25(a)

surface by a polyhedron of N faces, each of which is tangent to S at some point [Figure III–25(b)]. Note that this will automatically create a polygon [marked P in Figure III–25(b)] which is

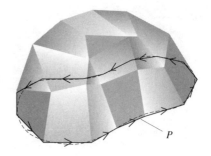

Figure III–25(b)

an approximation to the curve C. Let $\mathbf{F}(x, y, z)$ be a well-behaved vector function defined throughout the region of space occupied by the curve C and its capping surface S. Let us form the circulation of $\mathbf{F}$ around C_l, the boundary of the lth face of the polyhedron:

$$\oint_{C_l} \mathbf{F} \cdot \hat{\mathbf{t}} \, ds.$$

If we do this for each of the faces of the polyhedron and then add together all the circulations, we assert that this sum will be equal to the circulation of $\mathbf{F}$ around the polygon P:

$$\sum_{l=1}^{N} \oint_{C_l} \mathbf{F} \cdot \hat{\mathbf{t}} \, ds = \oint_{P} \mathbf{F} \cdot \hat{\mathbf{t}} \, ds. \qquad \text{(III–9)}$$

This is easy to prove. Consider two adjacent faces as shown in Figure III–26. The circulation about the face on the left [Figure III–26(a)] includes a term from the segment AB, which is

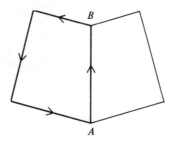

Figure III–26(a)

$\int_{A}^{B} \mathbf{F} \cdot \hat{\mathbf{t}} \, ds$. But the segment AB is common to both faces, and its contribution to the circulation around the *right*-hand face [Figure III–26(b)] is

$$\int_B^A \mathbf{F} \cdot \hat{\mathbf{t}} \, ds = - \int_A^B \mathbf{F} \cdot \hat{\mathbf{t}} \, ds.$$

We see that we traverse the common segment AB one way as part of the boundary of the left-hand face, and the other way as part of the boundary of the right-hand face. Thus, when we add

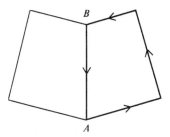

Figure III–26(b)

the circulations of $\mathbf{F}$ over the two faces, the segment AB contributes

$$\int_A^B \mathbf{F} \cdot \hat{\mathbf{t}} \, ds + \int_B^A \mathbf{F} \cdot \hat{\mathbf{t}} \, ds = 0.$$

It is clear that any segment common to two adjacent faces contributes nothing to the sum in Equation (III–9) because such segments always give rise to pairs of cancelling terms. But *all* segments are common to pairs of adjacent faces except those which, taken together, constitute the polygon P. This establishes Equation (III–9).

Now we go through an analysis very similar to that which yielded the divergence theorem. We write

$$\oint_P \mathbf{F} \cdot \hat{\mathbf{t}} \, ds = \sum_{l=1}^N \oint_{C_l} \mathbf{F} \cdot \hat{\mathbf{t}} \, ds$$

(III–10)

$$= \sum_{l=1}^N \left[\frac{1}{\Delta S_l} \oint_{C_l} \mathbf{F} \cdot \hat{\mathbf{t}} \, ds \right] \Delta S_l,$$

where ΔS_l is the area of the lth face. The quantity in the square brackets is, approximately, equal to $\hat{\mathbf{n}}_l \cdot (\nabla \times \mathbf{F})_l$ where $\hat{\mathbf{n}}_l$ is the unit positive normal on the lth face and $(\nabla \times \mathbf{F})_l$ is the curl of the vector function $\mathbf{F}$ evaluated at the point on the lth face at which it is tangent to S. We say "approximately" because it is actually the *limit* as ΔS_l tends to zero of the bracketed quantity in Equation (III–10), which is to be identified as $\hat{\mathbf{n}}_l \cdot (\nabla \times \mathbf{F})_l$. Ignoring this lack of rigor, we write

$$\lim_{\substack{N \to \infty \\ \text{each } \Delta S_l \to 0}} \sum_{l=1}^{N} \left[\frac{1}{\Delta S_l} \oint_{C_l} \mathbf{F} \cdot \hat{\mathbf{t}} \, ds \right] \Delta S_l$$

$$= \lim_{\substack{N \to \infty \\ \text{each } \Delta S_l \to 0}} \sum_{l=1}^{N} \hat{\mathbf{n}}_l \cdot (\nabla \times \mathbf{F})_l \, \Delta S_l$$

$$= \iint_{S} \hat{\mathbf{n}} \cdot \nabla \times \mathbf{F} \, dS. \tag{III–11}$$

Since the curve C is the limiting shape of the polygon P, we also have

$$\lim_{\substack{N \to \infty \\ \text{each} \Delta S_l \to 0}} \oint_{P} \mathbf{F} \cdot \hat{\mathbf{t}} \, ds = \oint_{C} \mathbf{F} \cdot \hat{\mathbf{t}} \, ds. \tag{III–12}$$

Combining Equations (III–10), (III–11), and (III–12), we arrive, finally, at Stokes' theorem:

$$\oint_{C} \mathbf{F} \cdot \hat{\mathbf{t}} \, ds = \iint_{S} \hat{\mathbf{n}} \cdot \nabla \times \mathbf{F} \, dS, \tag{III–13}$$

where S is *any* surface capping the curve C. Thus, in words, Stokes' theorem says that the line integral of the tangential component of a vector function over some closed path equals the surface integral of the normal component of the curl of that function integrated over any capping surface of the path. Stokes' theorem holds for any vector function $\mathbf{F}$ which is continuous and differentiable and has continuous derivatives on C and S.

Let's work an example. Take $\mathbf{F}(x, y, z) = \mathbf{i}z + \mathbf{j}x - \mathbf{k}x$, with C the circle of radius 1 centered at the origin and lying in the xy-plane, and S the part of the xy-plane enclosed by the circle [see Figure III–27(a)]. Now

$$\mathbf{F} \cdot \hat{\mathbf{t}} \, ds = z \, dx + x \, dy - x \, dz.$$

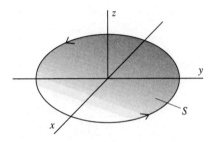

Figure III–27(a)

Thus, $\oint_C \mathbf{F} \cdot \hat{\mathbf{t}} \, ds = \oint x \, dy$. Heretofore we have always para-metrized curves with the arc length s. In this situation, however, the path C is most easily parametrized in terms of the angle θ shown in Figure III–27(b). Thus, we write

circulation

$$\oint x \, dy = \oint x \frac{dy}{d\theta} \, d\theta = \int_0^{2\pi} \cos^2 \theta \, d\theta = \pi, \quad \text{(III–14)}$$

where we use $x = \cos \theta$ and $y = \sin \theta$. Our next step is to notice

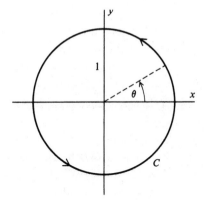

Figure III–27(b)

that the capping surface here is a portion of the xy-plane, so that the unit normal in the positive direction is $\hat{\mathbf{n}} = \mathbf{k}$. Thus,

$$\hat{\mathbf{n}} \cdot \nabla \times \mathbf{F} = \mathbf{k} \cdot \begin{vmatrix} \mathbf{i} & \mathbf{j} & \mathbf{k} \\ \partial/\partial x & \partial/\partial y & \partial/\partial z \\ z & x & -x \end{vmatrix} = 1,$$

and

$$\iint_S \hat{\mathbf{n}} \cdot \nabla \times \mathbf{F} \, dS = \iint_S dS = \pi, \qquad \text{(III–15)}$$

where this last equality follows from the fact that the surface integral in this case is merely the area of the unit circle. Since this result [Equation (III–15)] is identical with the one obtained above [Equation (III–14)], we have illustrated Stokes' theorem [Equation (III–13)]. As an exercise you should verify that the same result comes about from integrating $\hat{\mathbf{n}} \cdot \nabla \times \mathbf{F}$ over the hemisphere of radius 1 which also caps the curve C.

An Application of Stokes' Theorem

An important application of Stokes' theorem is provided by Ampère's circuital law. Consider any closed loop C enclosing a current I as in Figure III–28. Note that the direction of C and that

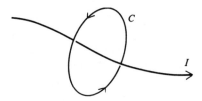

Figure III–28

of I correspond to the same right-hand rule which relates the directions of C and the positive normal to a surface capping C. Ampère's circuital law says that the line integral of the *magnetic field* $\mathbf{B}$ is related to the current thus:

$$\oint_C \mathbf{B} \cdot \hat{\mathbf{t}} \, ds = \mu_0 I$$

where the constant μ_0, called the permeability of free space, has the value 1.257×10^{-6} newtons per ampere^2. This law, like Gauss' law and the circulation law, says something about the integral of a field (the magnetic field in this case), and just as in the two previous cases, it is convenient to re-express it so that it will tell us something about the field at a point. To this end, we

An Application of first introduce the current density $\mathbf{J}$ (see page 52). Thus, if current
Stokes' Theorem is flowing through an area ΔS with normal $\hat{\mathbf{n}}$ (Figure III–29), the

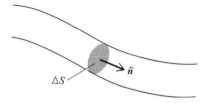

Figure III–29

current density $\mathbf{J}$ is such that

$$\Delta I = \mathbf{J} \cdot \hat{\mathbf{n}} \, \Delta S,$$

where ΔI is the total current. That is, current density is a vector
function whose magnitude is the current per unit area and whose
direction is that of the current flow. If $\mathbf{J}(x, y, z)$ is the current
density, then the total current flowing through a surface S is

$$\iint_S \mathbf{J} \cdot \hat{\mathbf{n}} \, dS.$$

Thus, Ampère's law can be written

$$\oint_C \mathbf{B} \cdot \hat{\mathbf{t}} \, ds = \mu_0 \iint_S \mathbf{J} \cdot \hat{\mathbf{n}} \, dS.$$

S can be any surface capping the curve C. If, as is usually the
case, the current flows through a wire the cross section of which
does not include the entire capping surface, it does not matter;
we can integrate over more than the wire cross section if we
remember that $\mathbf{J} \neq 0$ for that part of the surface S cut by the wire
and $\mathbf{J} = 0$ for the rest (Figure III–30). Thus,

$$\underbrace{\iint \mathbf{J} \cdot \hat{\mathbf{n}} \, dS}_{\substack{\text{cross section} \\ \text{of wire}}} = \underbrace{\iint_S \mathbf{J} \cdot \hat{\mathbf{n}} \, dS.}_{\substack{\text{entire capping} \\ \text{surface } S}}$$

Now using Stokes' theorem [Equation (III–13)], we have

99

$$\oint_C \mathbf{B} \cdot \hat{\mathbf{t}} \, dS = \iint_S \hat{\mathbf{n}} \cdot \nabla \times \mathbf{B} \, dS = \mu_0 \iint_S \hat{\mathbf{n}} \cdot \mathbf{J} \, dS.$$

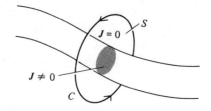

Figure III–30

Since C and S are arbitrary, we conclude that

$$\nabla \times \mathbf{B} = \mu_0 \mathbf{J}.$$

This is the differential form of Ampère's law. It is also a special case of one of Maxwell's equations, valid when the fields do not vary with time.

Stokes' Theorem and Simply Connected Regions

For many purposes, including some important applications, we must be able to assert that Stokes' theorem holds throughout some region D in three-dimensional space. By this we mean that we want the theorem to hold for any closed curve C lying entirely in D and any capping surface of C also lying entirely in D. This, of course, means the function $\mathbf{F}$ must be continuous and differentiable and have continuous first derivatives in D. But in addition we must impose a restriction on the region D itself. To understand how this comes about, suppose first that D is the interior of a sphere. If $\mathbf{F}$ is smooth[9] everywhere in D, then Stokes' theorem holds for any closed curve C lying entirely in D, and any capping surface of C also lying entirely in D. In other words, Stokes' theorem holds everywhere in D. A little thought should convince you that the same line of reasoning applies to the region between two concentric spheres provided $\mathbf{F}$ is smooth in that region. But for certain kinds of regions, troubles can arise. As an example, suppose D is the interior of a torus (roughly like a bagel or an inflated inner tube; see Figure III–31). The problem in this case is that it's possible to construct a closed curve in D like the

[9] Hereafter when we say that a function is "smooth," we'll mean that it is continuous, differentiable, and has continuous first derivatives.

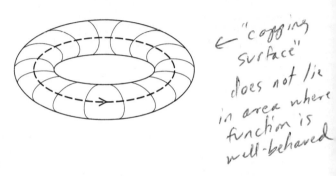

"capping surface" does not lie in area where function is well-behaved

Figure III–31

one shown in the figure with the property that none of its capping surfaces lies entirely in D. Although we insist that **F** be smooth in D, no conditions are imposed upon it elsewhere, so that outside the region it may not fulfill the requirements of smoothness that ensure the validity of Stokes' theorem. The relation between the line integral over C and the surface integral over S asserted by the theorem can, and in many cases does, break down if **F** is not smooth on S.

Mathematicians refer to regions such as the interior of a sphere or the space between two concentric spheres as *simply connected,* whereas the interior of a torus is not simply connected. By definition, a region D is simply connected if any closed curve lying entirely in D can shrink down to a point without leaving D. Using this definition, you should be able to verify that the interior of a sphere and the region between two concentric spheres are both simply connected, but that the interior of a torus is not. With the concept of simple connectedness available to us, we can easily specify the conditions under which Stokes' theorem holds throughout a region: The vector function **F** must be smooth everywhere in a *simply connected* region D. Then Stokes' theorem [Equation (III–13)] is valid for any closed curve C and any capping surface S of C, both of which lie entirely in D.

Most of the time we'll assume that the functions we work with are smooth and that the regions of interest are simply connected. There are situations, however, like the one discussed in the next section, where simple connectedness plays an essential role, and we'll point them out as we come to them.

Path Independence and the Curl

In our discussion of the differential form of the circulation law, we showed that because the line integral of an electrostatic field

E is zero over any closed path, the curl of E is zero. The same is true of any vector function F; that is, if

$$\oint_C \mathbf{F} \cdot \hat{\mathbf{t}} \, ds = 0$$

for all closed paths C, then

$$\nabla \times \mathbf{F} = 0.$$

The proof of this fact is precisely the same as the one given on pages 90–91 with E replaced everywhere by F.

Is the converse of this statement also true? That is, if $\nabla \times \mathbf{F} = 0$, does this imply that the circulation of F is zero over all closed paths? At first glance it might appear that the answer to this question is yes. All we have to do is use Stokes' theorem and observe that since by assumption $\nabla \times \mathbf{F} = 0$,

$$\oint_C \mathbf{F} \cdot \hat{\mathbf{t}} \, ds = \iint_S \hat{\mathbf{n}} \cdot \nabla \times \mathbf{F} \, dS = 0.$$

However, there is a flaw in this line of reasoning. Recall that the validity of Stokes' theorem requires that F be smooth *in a simply connected region*. If the region is not simply connected, Stokes' theorem may not hold, at least for some closed paths lying in the region, and the fact that $\nabla \times \mathbf{F} = 0$ does not guarantee that the circulation of F is zero over all closed paths. The closest we can come to a converse is to say that if $\nabla \times \mathbf{F} = 0$ *everywhere in a simply connected region,* then the circulation of F is zero for all closed paths in that region. The two statements "circulation equals zero" and "curl equals zero" are equivalent only in a simply connected region.

There is a slightly different, but often useful, way to state this connection between circulation and curl; namely, if $\int_C \mathbf{F} \cdot \hat{\mathbf{t}} \, ds$ is independent of path, then $\nabla \times \mathbf{F} = 0$, and if $\nabla \times \mathbf{F} = 0$ in a simply connected region, then $\int_C \mathbf{F} \cdot \hat{\mathbf{t}} \, ds$ is independent of path. You should have no difficulty in establishing this for yourself.

PROBLEMS

III–1 Use an argument like the one given in the text for the Coulomb force (pages 71–73) to show that $\int_C \mathbf{F} \cdot \hat{\mathbf{t}} \, ds$ is independent of path for *any* central force F.

Problems

III–2 In the text we obtained the result

$$(\nabla \times \mathbf{F})_z = \frac{\partial F_y}{\partial x} - \frac{\partial F_x}{\partial y}$$

by integrating over a small rectangular path. As an example of the fact that this result is independent of the path, rederive it, using the triangular path shown in the figure.

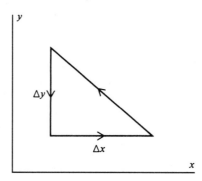

III–3 Calculate the curl of each of the following functions using Equation (III–7b):

(a) $\mathbf{i}z^2 + \mathbf{j}x^2 - \mathbf{k}y^2$.

(b) $3\mathbf{i}xz - \mathbf{k}x^2$.

(c) $\mathbf{i}e^{-y} + \mathbf{j}e^{-z} + \mathbf{k}e^{-x}$.

(d) $\mathbf{i}yz + \mathbf{j}xz + \mathbf{k}xy$.

(e) $-\mathbf{i}yz + \mathbf{j}xz$.

(f) $\mathbf{i}x + \mathbf{j}y + \mathbf{k}(x^2 + y^2)$.

(g) $\mathbf{i}xy + \mathbf{j}y^2 + \mathbf{k}yz$.

(h) $(\mathbf{i}x + \mathbf{j}y + \mathbf{k}z)/(x^2 + y^2 + z^2)^{3/2}$, $(x, y, z) \neq (0, 0, 0)$.

III–4 (a) Calculate $\oint \mathbf{F} \cdot \hat{\mathbf{t}}\, ds$ for the function in Problem III–3(a) over a square path of side s centered at $(x_0, y_0, 0)$, lying in the xy-plane, and oriented so that each side is parallel to the x- or y-axis.

(b) Divide the result of part (a) by the area of the square and take the limit of the quotient as $s \to 0$. Compare your result with the z-component of the curl found in Problem III–3(a).

(c) Repeat parts (a) and (b) for the functions in Problem III–3(b), (c), and (d). (You may find it interesting to try paths of different orientations and/or shapes.)

III–5 (a) Calculate $\oint \mathbf{F} \cdot \hat{\mathbf{t}}\, ds$ where

$$\mathbf{F} = \mathbf{k}(y + y^2)$$

over the perimeter of the triangle shown in the figure (integrate in the direction indicated by the arrows).

103

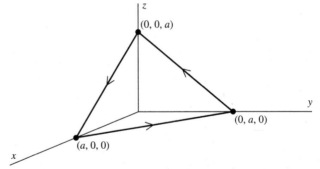

(b) Divide the result of part (a) by the area of the triangle and take the limit as $a \to 0$.

(c) Show that the result of part (b) is $\hat{\mathbf{n}} \cdot \nabla \times \mathbf{F}$ evaluated at $(0, 0, 0)$ where $\hat{\mathbf{n}}$ is the unit vector normal to the triangle and directed away from the origin.

III–6 Show that

$$\nabla \times \frac{\mathbf{A} \times \mathbf{r}}{2} = \mathbf{A},$$

where $\mathbf{r} = \mathbf{i}x + \mathbf{j}y + \mathbf{k}z$ and $\mathbf{A}$ is a constant vector.

III–7 Show that $\nabla \cdot (\nabla \times \mathbf{F}) = 0$. (Assume that mixed second partial derivatives are independent of the order of differentiation. For example, $\partial^2 F_z / \partial x\, \partial z = \partial^2 F_z / \partial z\, \partial x$.)

III–8 In the text (pages 82–84) we obtained the z-component of $\nabla \times \mathbf{F}$ in cylindrical coordinates. Proceeding the same way, obtain the θ- and r-components given on page 85.

III–9 Following the procedure suggested in the text (pages 82–85), obtain the expression for $\nabla \times \mathbf{F}$ in spherical coordinates given on page 85. The figures given on page 105 will be helpful.

III–10 (a) Rewrite the function in Problem III–3(e) in cylindrical coordinates and compute its curl using the expression given on page 85. Convert your result back to Cartesian coordinates and compare with the answer obtained in Problem III–3(e) (see Problem II–16).

(b) Repeat the above calculation for the function of Problem III–3(f).

III–11 (a) Rewrite the function in Problem III–3(g) in spherical coordinates and compute its curl using the expression given on page 85. Convert your result back to Cartesian coordinates and compare with the answer obtained in Problem III–3(g) (see Problem II–17).

(b) Repeat the above calculation for the function of Problem III–3(h).

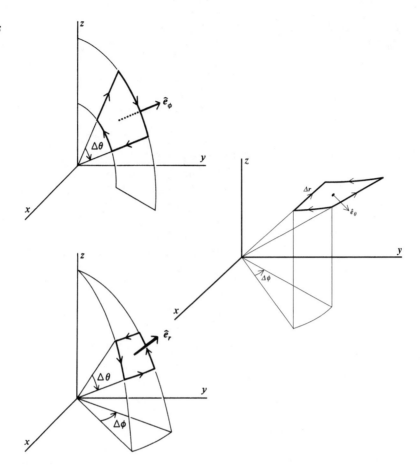

III–12 Any central force can be written in the form

$$\mathbf{F}(r) = \hat{\mathbf{e}}_r f(r),$$

where $\hat{\mathbf{e}}_r$ is a unit vector in the radial direction and f is a scalar function. Show by direct calculation of the curl that this function is irrotational (that is, $\nabla \times \mathbf{F} = 0$).

III–13 Which of the functions in Problem III–3 could be electrostatic fields?

III–14 Use Stokes' theorem to show that

$$\oint_C \hat{\mathbf{t}} \, ds = 0,$$

where C is a closed curve and $\hat{\mathbf{t}}$ is a unit vector tangent to the curve C.

III–15 Verify Stokes' theorem

$$\oint_C \mathbf{F} \cdot \hat{\mathbf{t}} \, ds = \iint_S \hat{\mathbf{n}} \cdot \nabla \times \mathbf{F} \, dS$$

in each of the following cases:

(a) $\mathbf{F} = \mathbf{i}z^2 - \mathbf{j}y^2$.

C, the square of side 1 lying in the xz-plane and directed as shown.

S, the five squares S_1, S_2, S_3, S_4, and S_5 as shown in the figure.

(b) $\mathbf{F} = \mathbf{i}y + \mathbf{j}z + \mathbf{k}x$.

C, the three quarter circle arcs C_1, C_2, and C_3 directed as shown in the figure.

S, the octant of the sphere $x^2 + y^2 + z^2 = 1$ enclosed by the three arcs.

(c) $\mathbf{F} = \mathbf{i}y - \mathbf{j}x + \mathbf{k}z$.

C, the circle of radius R lying in the xy-plane, centered at $(0, 0, 0)$ and directed as shown in the figure.

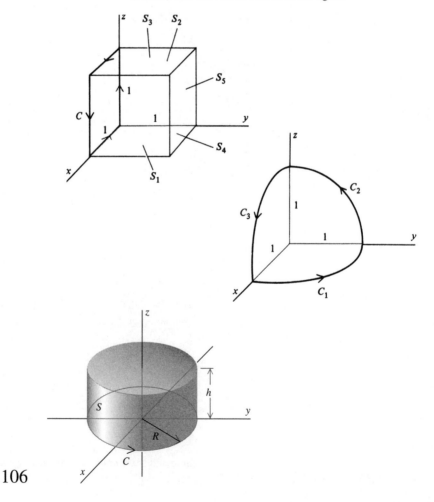

S, the curved and upper surfaces of the cylinder of radius R and height h.

III–16 (a) Consider a vector function with the property $\nabla \times \mathbf{F} = 0$ everywhere on two closed curves C_1 and C_2 and on any capping surface S of the region enclosed by them (see the figure). Show that the circulation of $\mathbf{F}$ around C_1 equals the circulation of $\mathbf{F}$ around C_2. In calculating the circulations direct the curves as indicated by the arrows in the figure.

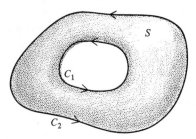

(b) The magnetic field due to an infinitely long straight wire carrying a uniform current I is $\mathbf{B} = (\mu_0 I / 2\pi r)\hat{\mathbf{e}}_\theta$. Show that $\nabla \times \mathbf{B} = 0$ everywhere except at $r = 0$.

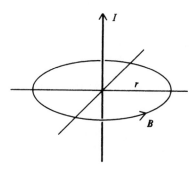

(c) Prove Ampère's circuital law for the field of the wire given in part (b). [*Hint:* Use the result of (b) to find the circulation of $\mathbf{B}$ around a circle with the wire passing through its center and normal to its plane. Then use the result of part (a) to relate this circulation to the circulation around an arbitrary curve enclosing the current.]

III–17 (a) Consider the function given in cylindrical coordinates by

$$\mathbf{F}(r, \theta, z) = \frac{\hat{\mathbf{e}}_\theta}{r}.$$

Show that Stokes' theorem does not hold for this function if C is the circle of radius R in the xy-plane centered at the origin, and S is the portion of the xy-plane enclosed by C. Why does the theorem fail in this case?

(b) Consider the region D which consists of all of three-dimensional space with the z-axis removed. Is the function $\mathbf{F}$ defined in (a) smooth in D? Does Stokes' theorem hold in D? Is D a simply connected region?

III–18 The electromotive force $\mathscr{E}$ in a circuit C is equal to the circulation of the electric field $\mathbf{E}$ around the circuit:

$$\mathscr{E} = \oint_C \mathbf{E} \cdot \hat{\mathbf{t}} \, ds.$$

Faraday discovered that in a stationary circuit an electromotive force is induced by a changing magnetic flux. That is,

$$\mathscr{E} = -\frac{d\Phi}{dt},$$

where

$$\Phi = \iint_S \mathbf{B} \cdot \hat{\mathbf{n}} \, dS,$$

t is time (don't confuse it with the tangent vector $\hat{\mathbf{t}}$), and S is any capping surface of C. Use this information and Stokes' theorem to derive the equation

$$\nabla \times \mathbf{E} = -\frac{\partial \mathbf{B}}{\partial t},$$

which is one of Maxwell's equations.

III–19 Determine the value of the line integral $\int_C \mathbf{F} \cdot \hat{\mathbf{t}} \, ds$ where

$$\mathbf{F} = (e^{-y} - ze^{-x})\mathbf{i} + (e^{-z} - xe^{-y})\mathbf{j} + (e^{-x} - ye^{-z})\mathbf{k}$$

and C is the path

$$\left.\begin{array}{c} x = \dfrac{1}{\ln 2} \ln(1+p), \\[2mm] y = \sin \dfrac{\pi p}{2}, \\[2mm] z = \dfrac{1 - e^p}{1 - e}, \end{array}\right\} \quad 0 \le p \le 1$$

from $(0, 0, 0)$ to $(1, 1, 1)$. [*Suggestion:* Think before you write!]

III–20 Maxwell's equations are

$$\nabla \cdot \mathbf{E} = \rho/\epsilon_0, \qquad\qquad \nabla \cdot \mathbf{B} = 0,$$

$$\nabla \times \mathbf{E} = -\frac{\partial \mathbf{B}}{\partial t}, \quad \text{and} \quad \nabla \times \mathbf{B} = \epsilon_0\mu_0 \frac{\partial \mathbf{E}}{\partial t} + \mu_0 \mathbf{J},$$

where **E** is the electric field, **B** the magnetic field, ρ the charge density, and **J** the current density. Use Maxwell's equations to derive the continuity equation

$$\nabla \cdot \mathbf{J} + \frac{\partial \rho}{\partial t} = 0.$$

Interpret this equation.

III–21 The electromagnetic field stores energy, and it is possible to show that in a volume V the amount of electromagnetic energy is

$$\iiint_V \rho_E \, dV,$$

where the energy density

$$\rho_E = \tfrac{1}{2}\,(\epsilon_0 \mathbf{E} \cdot \mathbf{E} + \mathbf{B} \cdot \mathbf{B}/\mu_0) = \tfrac{1}{2}\,(\epsilon_0 E^2 + B^2/\mu_0).$$

Use Maxwell's equations (see Problem III–20) to show that

$$\frac{\partial \rho_E}{\partial t} + \nabla \cdot \left(\frac{\mathbf{E} \times \mathbf{B}}{\mu_0} \right) = -\mathbf{J} \cdot \mathbf{E}.$$

Interpret this equation.

III–22 (a) Apply the divergence theorem to the function

$$\mathbf{G}(x, y) = \mathbf{i}G_x(x, y) + \mathbf{j}G_y(x, y),$$

using for V and S the volume and surface shown in the diagram; its bottom is a region R of the xy-plane, its top has the same shape as, and is parallel to, the bottom, and its side is parallel to the z-axis. In this way obtain the relation

$$\oint_C G_x \, dy - G_y \, dx = \iint_R \left(\frac{\partial G_x}{\partial x} + \frac{\partial G_y}{\partial y} \right) dx \, dy,$$

which is the divergence theorem in two dimensions.

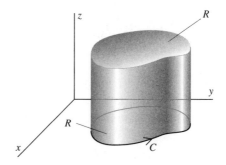

(b) Apply Stokes' theorem to the function

$$\mathbf{F}(x, y) = \mathbf{i}F_x(x, y) + \mathbf{j}F_y(x, y)$$

using for C a closed curve lying entirely in the xy-plane and for

S the region R of the xy-plane enclosed by C. In this way obtain the relation

$$\oint_C F_x \, dx + F_y \, dy = \iint_R \left(\frac{\partial F_y}{\partial x} - \frac{\partial F_x}{\partial y} \right) dx \, dy,$$

which is Stokes' theorem in two dimensions.

(c) Show that in two dimensions the divergence theorem and Stokes' theorem are identical.

III–23 (a) Let C be a closed curve lying in the xy-plane. What condition must the function $\mathbf{F}$ satisfy in order that

$$\oint_C \mathbf{F} \cdot \hat{\mathbf{t}} \, ds = A,$$

where A is the area enclosed by C? [*Hint:* See Problem III–22.]

(b) Give some examples of functions $\mathbf{F}$ having the property described in (a).

(c) Use line integrals to find formulas for the area of:
  (i) a rectangle.
  (ii) a right triangle.
  (iii) a circle.

(d) Show that the area enclosed by the plane curve C is the magnitude of

$$\frac{1}{2} \oint_C \mathbf{r} \times \hat{\mathbf{t}} \, ds,$$

where $\mathbf{r} = \mathbf{i}x + \mathbf{j}y$.

III–24 (a) There is an important theorem in vector calculus which says $\nabla \cdot \mathbf{G} = 0$ (where $\mathbf{G}$ is some differentiable vector function) implies and is implied by $\mathbf{G} = \nabla \times \mathbf{H}$ (where $\mathbf{H}$ is another differentiable function). To prove this we note first of all that $\mathbf{G} = \nabla \times \mathbf{H}$ implies that $\nabla \cdot \mathbf{G} = 0$ (see Problem III–7). To show that $\nabla \cdot \mathbf{G} = 0$ implies that we can write $\mathbf{G} = \nabla \times \mathbf{H}$, the simplest procedure is to give $\mathbf{H}$:

$H_x = 0,$

$$H_y = \int_{x_0}^{x} G_z(x', y, z) \, dx',$$

$$H_z = -\int_{x_0}^{x} G_y(x', y, z) \, dx' + \int_{y_0}^{y} G_x(x_0, y', z) dy',$$

where x_0 and y_0 are arbitrary constants. Show by direct calculation that if $\nabla \cdot \mathbf{G} = 0$, then $\mathbf{G} = \nabla \times \mathbf{H}$.

(b) Is the vector function $\mathbf{H}$ specified in (a) unique? That is, can we alter it in any way without invalidating the relation $\mathbf{G} = \nabla \times \mathbf{H}$?

III–25 Determine in which of the following cases it is possible to write $\mathbf{G} = \nabla \times \mathbf{H}$. In the cases where it is possible, find $\mathbf{H}$ (see Problem III–24).

(a) $\mathbf{G} = \mathbf{i}y + \mathbf{j}z + \mathbf{k}x$.
(b) $\mathbf{G} = B_0\mathbf{k}$, B_0 a constant.
(c) $\mathbf{G} = \mathbf{i}x^2 - \mathbf{k}y^2$.
(d) $\mathbf{G} = 2\mathbf{i}x - \mathbf{j}y - \mathbf{k}z$.
(e) $\mathbf{G} = 2\mathbf{i}x - \mathbf{j}y + \mathbf{k}z$.

III–26 Since the divergence of any magnetic field $\mathbf{B}$ is zero, we can write $\mathbf{B} = \nabla \times \mathbf{A}$ (see Problem III–24). Prove that the circulation of $\mathbf{A}$ around an arbitrary closed path C is equal to the flux of $\mathbf{B}$ through any surface S capping C.

III–27 Prove the statement made in Problem III–24(a) by applying Stokes' theorem and the divergence theorem. [*Hint:* See the diagram below.]

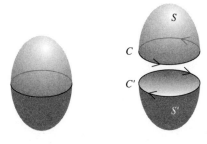

III–28 (a) What is the integral form of the equation $\mathbf{G} = \nabla \times \mathbf{H}$? [*Hint:* Compare the differential and integral forms of Ampère's circuital law.]
(b) Verify your result in part (a) using for $\mathbf{G}$ and $\mathbf{H}$ functions selected from Problem III–25, and paths and surfaces of integration of your own choice.

III–29 In the text we defined the curl as the limit of a certain ratio. An alternative definition is provided by the equation

$$\nabla \times \mathbf{F} = \lim_{\Delta V \to 0} \frac{1}{\Delta V} \iint_S \hat{\mathbf{n}} \times \mathbf{F} \, dS,$$

where $\mathbf{F}$ is a vector function of position, the integration is carried out over a closed surface S which enclosed the volume ΔV, and $\hat{\mathbf{n}}$ is the unit vector normal to S pointing outward from the enclosed volume.

(This definition does not display the geometric significance of the curl as well as the one given in the text. Nonetheless, in one respect at least it may be preferable; it gives the $\nabla \times \mathbf{F}$ rather than just a component of it.)

(a) Following a procedure similar to the one used in the text in treating the divergence, integrate over a "cuboid" and show that the definition given above yields Equation (III–7b).

(b) Arguing as we did in the text in establishing the divergence theorem, use the above expression for the curl to derive the equation

$$\iint_S \hat{\mathbf{n}} \times \mathbf{F} \; dS = \iiint_V \nabla \times \mathbf{F} \; dV,$$

where V is the volume enclosed by S.

(c) Derive the equation of part (b) directly from the divergence theorem. [*Hint:* In the divergence theorem [Equation (II–30)] replace $\mathbf{F}$ by $\mathbf{e} \times \mathbf{F}$ where $\mathbf{e}$ is an arbitrary constant vector.]

(d) Verify the equation of part (b) for $\mathbf{F} = \mathbf{i}y - \mathbf{j}z + \mathbf{k}x$ and V the unit cube shown in the figure.

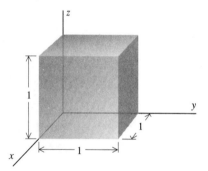

III–30 The result

$$(\nabla \times \mathbf{F})_z = \frac{\partial F_y}{\partial x} - \frac{\partial F_x}{\partial y}$$

has been established by calculating the circulation of $\mathbf{F}$ around a rectangle (see the text, pages 75 ff.) and around a right triangle (see Problem III–2). In this problem you will show that the result holds when the circulation is calculated around *any* closed curve lying in the *xy*-plane.

(a) Approximate an arbitrary closed curve C in the *xy*-plane by a polygon P as shown in the figure. Subdivide the area enclosed by P into N patches of which the *l*th has area ΔS_l. Convince yourself by means of a sketch that this subdivision can be made with only two kinds of patches: rectangles and right triangles.

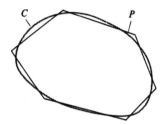

(b) Letting $C(x, y) = \partial F_y/\partial x - \partial F_x/\partial y$, use Taylor series to show that for N large and each ΔS_l small,

$$\oint_P \mathbf{F} \cdot \hat{\mathbf{t}} \, ds = \sum_{l=1}^{N} \oint_{C_l} \mathbf{F} \cdot \hat{\mathbf{t}} \, ds$$

$$\cong C(x_0, y_0) \, \Delta A + \left(\frac{\partial C}{\partial x}\right)_{x_0, y_0} \sum_{l=1}^{N} (x_l - x_0) \, \Delta S_l$$

$$+ \left(\frac{\partial C}{\partial y}\right)_{x_0, y_0} \sum_{l=1}^{N} (y_l - y_0) \, \Delta S_l + \cdots,$$

where C_l is the perimeter of the lth patch, (x_0, y_0) is some point in the region enclosed by P, and ΔA is the area enclosed by P.

(c) Show that

$$\lim_{\substack{N \to \infty \\ \text{each } \Delta S_l \to 0}} \oint_P \mathbf{F} \cdot \hat{\mathbf{t}} \, ds = \oint_C \mathbf{F} \cdot \hat{\mathbf{t}} \, ds$$

$$= \left[C(x_0, y_0) + (\bar{x} - x_0) \left(\frac{\partial C}{\partial x}\right)_{x_0, y_0}\right.$$

$$\left. + (\bar{y} - y_0) \left(\frac{\partial C}{\partial y}\right)_{x_0, y_0} + \cdots \right] \Delta S,$$

where ΔS is the area of the region R enclosed by C and $(\bar{x}, \bar{y})$ are the coordinates of the centroid of the region R; that is,

$$\bar{x} = \frac{1}{\Delta S} \iint_R x \, dx \, dy \quad \text{and} \quad \bar{y} = \frac{1}{\Delta S} \iint_R y \, dx \, dy.$$

(d) Finally, calculate

$$(\nabla \times \mathbf{F})_z = \lim_{\substack{\Delta S \to 0 \\ \text{about } x_0, y_0}} \frac{1}{\Delta S} \oint_C \mathbf{F} \cdot \hat{\mathbf{t}} \, ds.$$

Chapter IV

The Gradient

For mostly they goes up and down . . .

P. R. Chalmers

Line Integrals and the Gradient

We have now investigated the relationship between the following two statements:

1. $\oint_C \mathbf{F} \cdot \hat{\mathbf{t}} \, ds = 0$ for any closed curve C.
2. $\nabla \times \mathbf{F} = 0$.

We saw in the last chapter that the first of these statements implies the second and is equivalent to the assertion that the line integral of $\mathbf{F} \cdot \hat{\mathbf{t}}$ is independent of the path. We also saw that the second statement implies the first if $\mathbf{F}$ is smooth in a simply connected region. You might think that two ways of saying something would be enough, but there is a *third* way, as we shall now see.

114

Let us suppose that a given vector function $\mathbf{F}(x, y, z)$ has associated with it a scalar function $\psi(x, y, z)$ and that the two functions are related as follows:

$$F_x = \frac{\partial\psi}{\partial x}, \qquad F_y = \frac{\partial\psi}{\partial y}, \qquad \text{and} \qquad F_z = \frac{\partial\psi}{\partial z}. \qquad \text{(IV-1)}$$

If the above relations hold, then the line integral of $\mathbf{F} \cdot \hat{\mathbf{t}}$ is independent of path. To show this, we use the three relations given in Equation (IV–1) and the formula for the unit tangent vector to get

$$\mathbf{F} \cdot \hat{\mathbf{t}} = \frac{\partial\psi}{\partial x}\frac{dx}{ds} + \frac{\partial\psi}{\partial y}\frac{dy}{ds} + \frac{\partial\psi}{\partial z}\frac{dz}{ds} = \frac{d\psi}{ds}$$

where the second equality follows from a familiar chain rule of multivariate calculus. Suppose now that the path C joins the two points (x_0, y_0, z_0) and (x_1, y_1, z_1). Then

$$\int_C \mathbf{F} \cdot \hat{\mathbf{t}}\, ds = \int_C \frac{d\psi}{ds}\, ds = \int_C d\psi$$

$$= \psi(x_1, y_1, z_1) - \psi(x_0, y_0, z_0).$$

You can see that this result depends only on the points at which the path C begins and ends. We'd get the same result for *any* path joining these two points. This proves our assertion: with $\mathbf{F}$ and ψ related as in Equations (IV–1), the line integral of $\mathbf{F} \cdot \hat{\mathbf{t}}$ is independent of path. We shall now show that the converse of this statement is also true; that is, if the line integral of $\mathbf{F} \cdot \hat{\mathbf{t}}$ is independent of path, there is a scalar function $\psi(x, y, z)$ related to $\mathbf{F}$ as specified in Equations (IV–1).

We begin with the observation that, because the line integral $\int_C \mathbf{F} \cdot \hat{\mathbf{t}}\, ds$ is independent of path, if we integrate from some *fixed* point $P_0(x_0, y_0, z_0)$ to a second point $P(x, y, z)$, the result is a scalar function of the coordinates (x, y, z):

$$\psi(x, y, z) = \int_{(x_0, y_0, z_0)}^{(x, y, z)} \mathbf{F} \cdot \hat{\mathbf{t}}\, ds. \qquad \text{(IV-2)}$$

It is important to understand that this would not be true if the integral depended on path, for then its value would depend not only on the coordinates (x, y, z) of the point P but also on the path joining P_0 and P, and the integral would not then be a function within the standard definition of the term.

Since the integral we're examining is path-independent, we are free to select any curve as the path of integration. We choose the one shown in Figure IV–1. It consists of two parts. The first, C_0,

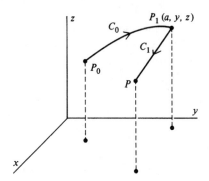

Figure IV–1

connects P_0 to an intermediate point P_1 whose coordinates are (a, y, z) where a is some constant. Beyond fixing its two end points and requiring it to be reasonably smooth, we do not need to specify anything more about C_0. The second part of the curve, C_1, is the straight line segment from P_1 to P. Thus, Equation (IV–2) becomes

$$\psi(x, y, z) = \int_{P_0}^{P_1} \mathbf{F} \cdot \hat{\mathbf{t}} \, ds + \int_{P_1}^{P} F_x(x', y, z) \, dx'.$$

The first term on the right-hand side of this equation is independent of the variable x. The second term is, effectively, nothing more than an ordinary one-dimensional integral since y and z are constant on C_1 and just come along for the ride. That is,

$$\int_{P_1}^{P} F_x(x', y, z) \, dx' = \int_{a}^{x} F_x(x', y = \text{const.}, z = \text{const.}) \, dx',$$

and so

$$\frac{\partial \psi}{\partial x} = \frac{d}{dx} \int_{a}^{x} F_x(x', y = \text{const.}, z = \text{const.}) \, dx'$$

$$= F_x(x, y, z),$$

where we use the fact that the derivative of an integral with respect to its upper limit is merely the integrand evaluated at that limit. This establishes one of the three relations we sought. The other two, $F_y = \partial\psi/\partial y$ and $F_z = \partial\psi/\partial z$, can be obtained by the same sort of reasoning, and you should carry out the derivations yourself. Figure IV–2(a) and (b) will be helpful.

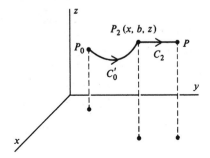

Figure IV–2(a)

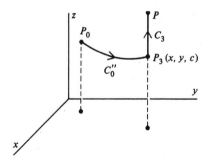

Figure IV–2(b)

You have probably recognized by now that we have here another use for the del notation. That is,

$$F_x = \frac{\partial\psi}{\partial x}, \qquad F_y = \frac{\partial\psi}{\partial y}, \qquad \text{and} \qquad F_z = \frac{\partial\psi}{\partial z}$$

can be combined to give

$$\mathbf{F} = \mathbf{i}\,\frac{\partial\psi}{\partial x} + \mathbf{j}\,\frac{\partial\psi}{\partial y} + \mathbf{k}\,\frac{\partial\psi}{\partial z}$$

$$= \left(\mathbf{i}\,\frac{\partial}{\partial x} + \mathbf{j}\,\frac{\partial}{\partial y} + \mathbf{k}\,\frac{\partial}{\partial z}\right)\psi = \nabla\psi,$$

117

which is read "del psi." This operator is called the *gradient* and is sometimes written gradψ. However, we shall always write ∇ψ in keeping with modern usage. The gradient of ψ is a *vector function* of position. Its geometric significance will be discussed in detail below.

We have now established the relationship between path independence and the existence of a scalar function $\psi(x, y, z)$ such that $\mathbf{F} = \nabla\psi$. Since there is also a relationship between path independence and the fact that $\nabla \times \mathbf{F} = 0$, you may suspect that $\nabla \times \mathbf{F} = 0$ and $\mathbf{F} = \nabla\psi$ are also related. Indeed, if $\mathbf{F} = \nabla\psi$, then under suitable conditions, $\nabla \times \mathbf{F} = 0$. This is easily established. Consider, for example, the x-component of $\nabla \times \mathbf{F}$:

$$(\nabla \times \mathbf{F})_x = \frac{\partial F_z}{\partial y} - \frac{\partial F_y}{\partial z} = \frac{\partial}{\partial y}\left(\frac{\partial\psi}{\partial z}\right) - \frac{\partial}{\partial z}\left(\frac{\partial\psi}{\partial y}\right)$$

$$= \frac{\partial^2\psi}{\partial y\,\partial z} - \frac{\partial^2\psi}{\partial z\,\partial y} = 0.$$

This last equality follows if ψ and its first and second derivatives are continuous for then $\partial^2\psi/\partial y\,\partial z = \partial^2\psi/\partial z\,\partial y$. Obviously the other two components of $\nabla \times \mathbf{F}$ can be shown to vanish in exactly the same way. Thus,

$$F_q = \frac{\partial\psi}{\partial q} \quad (q = x, y, z) \quad \Rightarrow \quad \nabla \times \mathbf{F} = 0.$$

The converse of what we have just shown would assert that if $\nabla \times \mathbf{F} = 0$, then there exists a scalar function ψ such that $\mathbf{F} = \nabla\psi$, a statement that is true provided the region of interest is simply connected. To understand this, we can consult Figure IV–3, which shows how path independence of the line integral of $\mathbf{F} \cdot \hat{\mathbf{t}}$, $\nabla \times \mathbf{F} = 0$, and $\mathbf{F} = \nabla\psi$ are related. The solid arrows in the diagram represent implications which hold in general provided $\mathbf{F}$ is smooth. The dashed arrows represent implications requiring not only that $\mathbf{F}$ be smooth, but that the region of interest be simply connected. We have already shown that (1) implies both (2) and (3) and that (3) implies (1) in a simply connected region. Combining these two statements, we see that (3) implies (2) in a simply connected region.

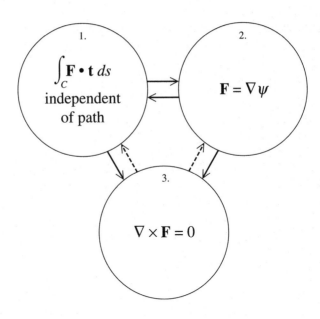

Figure IV–3

In practice, just as the functions we deal with usually have continuous first derivatives (and are therefore smooth), the regions we work with are simply connected. In such circumstances we can relax a bit and regard the three statements summarized in Figure IV–3 as equivalent: each implies and is implied by each of the others. However, you should be aware of simple connectedness and its implications for the relations among the three statements.

To give a simple example of the ideas we have been discussing, consider the vector function

$$\mathbf{F}(x, y, z) = \mathbf{i}y + \mathbf{j}x.$$

This function is smooth everywhere, and we have already noted that its curl is zero (page 91). According to what we have just said, this means there must be a scalar function $\psi(x, y, z)$ such that $\mathbf{F}$ is its gradient. Thus, ψ must satisfy

$$F_x = y = \frac{\partial \psi}{\partial x}, \qquad F_y = x = \frac{\partial \psi}{\partial y}, \qquad F_z = 0 = \frac{\partial \psi}{\partial z}.$$

119

Clearly $\psi(x, y, z) = xy + C$, where C is an arbitrary constant, satisfies these relations. This should be contrasted with the case of the function $\mathbf{F} = \mathbf{i}y - \mathbf{j}x$, the curl of which does *not* vanish (page 92). If this function were the gradient of a scalar function ψ, we should have

$$F_x = y = \frac{\partial \psi}{\partial x}, \qquad F_y = -x = \frac{\partial \psi}{\partial y}, \qquad F_z = 0 = \frac{\partial \psi}{\partial z},$$

but, as you should be able to convince yourself, there is no function ψ which satisfies these three equations.

The expression we have written for the gradient of a scalar function $\psi(x, y, z)$, namely

$$\nabla \psi = \mathbf{i} \frac{\partial \psi}{\partial x} + \mathbf{j} \frac{\partial \psi}{\partial y} + \mathbf{k} \frac{\partial \psi}{\partial z},$$

is really just the form of this operator in Cartesian coordinates. To find the form of the gradient in other coordinate systems, if you go about it straightforwardly, is a tedious job. For example, to find the gradient in cylindrical coordinates, we would first have to express the Cartesian unit vectors $\mathbf{i}$, $\mathbf{j}$, and $\mathbf{k}$ in terms of the analogous quantities $\hat{\mathbf{e}}_r$, $\hat{\mathbf{e}}_\theta$, and $\hat{\mathbf{e}}_z$ in cylindrical coordinates. Then, using $x = r \cos \theta$, $y = r \sin \theta$, and the chain rule for differentiation, we would have to express derivatives with respect to x, y, and z in terms of those with respect to r, θ, and z. We shall not pursue this matter here because later (see pages 140 ff.) an easier and faster method will be available to us. For the present we merely quote the form of the gradient in cylindrical and in spherical coordinates.

Cylindrical:

$$\nabla \psi = \hat{\mathbf{e}}_r \frac{\partial \psi}{\partial r} + \hat{\mathbf{e}}_\theta \frac{1}{r} \frac{\partial \psi}{\partial \theta} + \hat{\mathbf{e}}_z \frac{\partial \psi}{\partial z}. \qquad \text{(IV--3)}$$

Spherical:

$$\nabla \psi = \hat{\mathbf{e}}_r \frac{\partial \psi}{\partial r} + \hat{\mathbf{e}}_\theta \frac{1}{r} \frac{\partial \psi}{\partial \theta} + \hat{\mathbf{e}}_\phi \frac{1}{r \sin \theta} \frac{\partial \psi}{\partial \phi}. \qquad \text{(IV--4)}$$

A coordinate-free definition of the gradient analogous to the ones given for the divergence [Equation (II–17)] and the curl [Equation (III–8)] is discussed in Problem IV–25.

Finding the Electrostatic Field

We began our discussion of vector calculus with a search for some convenient method for finding the electrostatic field. Our investigations led us to the differential form of Gauss' law,

$$\nabla \cdot \mathbf{E} = \rho/\epsilon_0.$$

Even this expression is not often useful for finding $\mathbf{E}$ because it is one equation in three unknowns (E_x, E_y, and E_z in Cartesian coordinates). Now, at last, we are able to complete our discussion and write down the equations which are often the most useful of all known methods for finding the field.

This final step rests on the observation that since

$$\oint_C \mathbf{E} \cdot \hat{\mathbf{t}} \, ds = 0$$

for any closed path C, the field $\mathbf{E}$ can be written as the gradient of a scalar function. Conventionally this function, called the electrostatic potential, is designated $\Phi(x, y, z)$, and we write[1]

$$\mathbf{E} = -\nabla\Phi.$$

Combining this equation with the differential form of Gauss' law [Equation (II–17)], we get

$$\nabla \cdot (-\nabla\Phi) = \rho/\epsilon_0,$$

or

$$\nabla \cdot (\nabla\Phi) = -\rho/\epsilon_0.$$

---

[1] The negative sign in this equation is not put there just to make life more difficult; there is a good reason for it. See the discussion on pages 138–39.

When we write out the left-hand side of this equation in detail, we find

$$\nabla \cdot (\nabla \Phi) = \left(\mathbf{i}\, \frac{\partial}{\partial x} + \mathbf{j}\, \frac{\partial}{\partial y} + \mathbf{k}\, \frac{\partial}{\partial z} \right) \cdot \left(\mathbf{i}\, \frac{\partial \Phi}{\partial x} + \mathbf{j}\, \frac{\partial \Phi}{\partial y} + \mathbf{k}\, \frac{\partial \Phi}{\partial z} \right)$$

$$= \frac{\partial^2 \Phi}{\partial x^2} + \frac{\partial^2 \Phi}{\partial y^2} + \frac{\partial^2 \Phi}{\partial z^2} \, ,$$

and so

$$\frac{\partial^2 \Phi}{\partial x^2} + \frac{\partial^2 \Phi}{\partial y^2} + \frac{\partial^2 \Phi}{\partial z^2} = -\rho/\epsilon_0. \tag{IV-5}$$

Equation (IV–5) can be written more compactly by introducing a new operator, called the *Laplacian,* which is denoted, for fairly obvious reasons, by the symbol ∇^2 (read "del squared"). That is,

$$\nabla^2 = \nabla \cdot \nabla = \left(\mathbf{i}\, \frac{\partial}{\partial x} + \mathbf{j}\, \frac{\partial}{\partial y} + \mathbf{k}\, \frac{\partial}{\partial z} \right) \cdot \left(\mathbf{i}\, \frac{\partial}{\partial x} + \mathbf{j}\, \frac{\partial}{\partial y} + \mathbf{k}\, \frac{\partial}{\partial z} \right)$$

$$= \frac{\partial^2}{\partial x^2} + \frac{\partial^2}{\partial y^2} + \frac{\partial^2}{\partial z^2} \, . \tag{IV-6}$$

In this new notation Equation (IV–5) becomes

Poisson

$$\nabla^2 \Phi = -\rho/\epsilon_0. \tag{IV-7}$$

Equation (IV–6) provides the form of the Laplacian in Cartesian coordinates; its forms in cylindrical and spherical coordinates will be given in the next section. The best definition of the Laplacian is probably

$$\nabla^2 f = \nabla \cdot (\nabla f),$$

where f is some suitably continuous scalar function of position. This definition has the important advantage of being independent of the coordinate system.

Equation (IV–7) is called Poisson's equation. It is a linear, second-order partial differential equation in *one* unknown, the scalar function $\Phi(x, y, z)$, and is the culmination of our long search for a method of determining the electrostatic field. A great body of work exists describing the many elegant mathematical schemes which have been devised to solve it, and a few simple examples are given in the next section. In any problem, once we have Φ, the field is trivial to find using $\mathbf{E} = -\nabla\Phi$.

At any point in space where there is no electric charge, the density ρ is zero and Poisson's equation reduces to

$$\nabla^2\Phi = 0. \qquad \text{Laplace}$$

This is called *Laplace's equation* and is more often used than Poisson's equation. The reason for this is that usually charges are distributed over various objects; this gives rise to a field, and we are interested in finding the potential (and from it, the field) in the charge-free space between the objects. In the simplest of situations it is possible to specify "boundary conditions," that is, the value of the potential on the surfaces of these objects (Figure IV–4). We then find that solution of Laplace's equation

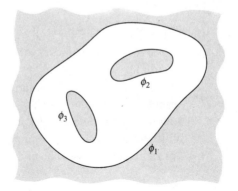

Figure IV–4

which takes on the given values on the surfaces. This is illustrated in the next section.

Using Laplace's Equation[2]

Whether solving Laplace's equation is or is not a topic in vector calculus is a moot point, but the basis of our entire discussion

[2] This section is not essential to what follows and may be omitted.

has been a search for a method to calculate electric fields. Since Laplace's equation is the end product of that search, we can scarcely omit a few examples to show how it works.

We begin with an especially simple problem. Imagine we have two very large (''infinite'') parallel plates separated by a distance s (Figure IV–5). Choosing a coordinate system as shown in the

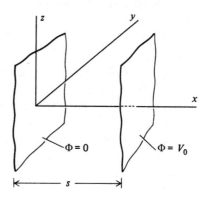

Figure IV–5

figure, let the plate at $x = 0$ be held at zero potential and that at $x = s$ at V_0. Our object is to find the potential and the electric field in the space between the two plates. Because the plates are infinitely large, there is nothing to distinguish a point (x, y, z) from any other point (x, y', z') having the same x-coordinate. It follows that the potential Φ depends on x but not on y or z. Thus, $\nabla^2\Phi$ reduces in this case simply to $d^2\Phi/dx^2$, and so Laplace's equation and the associated boundary conditions are

$$\frac{d^2\Phi}{dx^2} = 0$$

and

$$\Phi = \begin{cases} 0 & \text{at } x = 0 \\ V_0 & \text{at } x = s. \end{cases}$$

This is a trivial problem and the solution is

$$\Phi(x) = \frac{V_0 x}{s}.$$

The electric field is found using $\mathbf{E} = -\nabla\Phi$, which yields

$$E_x = \frac{-V_0}{s} \qquad \text{and} \qquad E_y = E_z = 0.$$

Thus, the field is a constant vector normal to the plates. This is an excellent approximation to the potential and field between, but far from the edges of, two plates whose linear dimensions are large compared with their separation. You may recognize this arrangement as a parallel plate capacitor.

Our second example is a spherical capacitor, that is, two con- centric spheres having radii R_1 and R_2 with the inner one maintained at a potential V_0 and the outer at zero (Figure IV–6). We

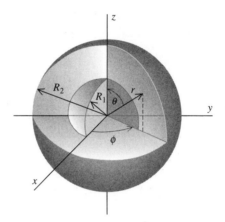

Figure IV–6

are required to find the potential and field everywhere between the spheres. In this situation we would obviously do well to work in spherical coordinates r, θ, and ϕ, in which Laplace's equation between the spheres has the imposing form

$$\nabla^2\Phi = \frac{1}{r^2}\frac{\partial}{\partial r}\left(r^2\frac{\partial\Phi}{\partial r}\right)$$

$$+ \frac{1}{r^2 \sin\theta}\frac{\partial}{\partial\theta}\left(\sin\theta\frac{\partial\Phi}{\partial\theta}\right) + \frac{1}{r^2 \sin^2\theta}\frac{\partial^2\Phi}{\partial\phi^2} = 0.$$

(See Problem IV–23.) Fortunately, we need not work with this equation as it stands; a little thought will convince you that Φ can only be a function of r since there is no way to distinguish a point (r, θ, ϕ) from another (r, θ', ϕ') with the same r but different θ and ϕ. Thus,

$$\frac{\partial \Phi}{\partial \theta} = \frac{\partial \Phi}{\partial \phi} = 0,$$

and Laplace's equation reduces to

$$\frac{1}{r^2} \frac{d}{dr}\left(r^2 \frac{d\Phi}{dr}\right) = 0. \tag{IV–8}$$

We are interested in the solution of this equation which is valid for $R_1 < r < R_2$ and satisfies the boundary conditions

$$\Phi(r) = \begin{cases} V_0 & \text{at } r = R_1 \\ 0 & \text{at } r = R_2. \end{cases}$$

Multiplying Equation (IV–8) by r^2 and putting $\psi = d\Phi/dr$, we get

$$\frac{d}{dr}(r^2\psi) = 0,$$

and so

$$r^2\psi = c_1,$$

where c_1 is a constant. Hence,

$$\psi = \frac{d\Phi}{dr} = \frac{c_1}{r^2},$$

and it follows that

$$\Phi = -\frac{c_1}{r} + c_2, \tag{IV–9}$$

where c_2 is another constant. Imposing the boundary conditions, we find

$$-\frac{c_1}{R_1} + c_2 = V_0 \quad \text{and} \quad -\frac{c_1}{R_2} + c_2 = 0,$$

whence

$$c_1 = \frac{V_0 R_1 R_2}{R_1 - R_2} \quad \text{and} \quad c_2 = \frac{V_0 R_1}{R_1 - R_2}.$$

Substituting these in the expression for the potential [Equation (IV–9)], we get

$$\Phi(r) = \frac{V_0 R_1}{R_1 - R_2} \left(1 - \frac{R_2}{r}\right), \quad R_1 < r < R_2.$$

To get the electric field, we must take the gradient of Φ, and this is clearly most conveniently done in spherical coordinates [see Equation (IV–4)]. However, since in this case Φ depends only on r, we get only a radial component:

$$E_r = -\frac{d\Phi}{dr} = -\frac{V_0 R_1 R_2}{R_1 - R_2}\frac{1}{r^2},$$

$$E_\theta = E_\Phi = 0, \quad (R_1 < r < R_2).$$

Our third and last example is more complicated (and more *ex 3* interesting) than the foregoing. If a potential difference is maintained between two "infinite" parallel plates P and P' (Figure IV–7), then we know from our first example that the field

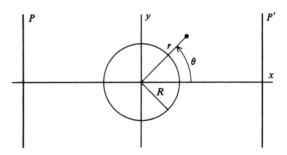

Figure IV–7

between them is a constant vector normal to the plates. Choosing a coordinate system as shown in the figure (with the z-axis out of the plane of the paper), we have $\mathbf{E} = E_0 \mathbf{i}$ where E_0 is a constant. Let an "infinitely" long cylinder held at zero potential be situated between the plates with its axis along the z-axis. Let its radius R be small compared with the plate separation. What are the potential and the electric field outside the cylinder and between the plates? Here, clearly, we should use cylindrical coordinates (r, θ, z), in which case Laplace's equation reads

$$\nabla^2\Phi = \frac{1}{r}\frac{\partial}{\partial r}\left(r\frac{\partial\Phi}{\partial r}\right) + \frac{1}{r^2}\frac{\partial^2\Phi}{\partial\theta^2} + \frac{\partial^2\Phi}{\partial z^2} = 0.$$

(See Problem IV–21.) You should convince yourself that Φ in this case must be independent of z, so this equation simplifies somewhat to

$$\frac{1}{r}\frac{\partial}{\partial r}\left(r\frac{\partial\Phi}{\partial r}\right) + \frac{1}{r^2}\frac{\partial^2\Phi}{\partial\theta^2} = 0. \tag{IV–10}$$

There are two boundary conditions of which the first is

$$\Phi(r, \theta) = 0 \quad \text{at} \quad r = R.$$

The second condition has to do with the fact that at large values of r, the influence of the cylinder is negligible and the field must be, to a good approximation, what it would be if the cylinder were not present at all, that is, $E_0\mathbf{i}$. To put this in terms of the potential, we note that

$$\Phi = -E_0 x$$

will provide just such a field. Since $x = r \cos \theta$, we can write the second boundary condition

$$\Phi(r, \theta) = -E_0 r \cos \theta, \quad r \gg R. \tag{IV–11}$$

Let's try to solve Laplace's equation for this problem [Equation (IV–10)] by assuming we can write

$$\Phi(r, \theta) = f(r) \cos \theta, \tag{IV–12}$$

where $f(r)$ is an as yet unknown function. What prompts us to do this is the fact that the second boundary condition [Equation (IV–11)] has precisely this form—a function of r multiplied by cos θ. If we substitute Equation (IV–12) into Equation (IV–10),

Using Laplace's Equation

the result is a differential equation for the function $f(r)$:

$$\frac{d^2f}{dr^2} + \frac{1}{r}\frac{df}{dr} - \frac{1}{r^2}f = 0.$$

Putting $f(r) = r^\lambda$ where λ is a constant leads to

$$\lambda(\lambda - 1)r^{\lambda-2} + \lambda r^{\lambda-2} - r^{\lambda-2} = 0,$$

or

$$\lambda^2 = 1,$$

and $\lambda = \pm 1$. Hence we get

$$f(r) = Ar + \frac{B}{r},$$

where A and B are constants. Thus, our solution is

$$\Phi(r, \theta) = \left(Ar + \frac{B}{r}\right)\cos\theta.$$

The first boundary condition requires that

$$AR + \frac{B}{R} = 0,$$

or

$$B = -AR^2.$$

Hence,

$$\Phi(r, \theta) = Ar\cos\theta - \frac{AR^2}{r}\cos\theta.$$

To impose the second condition, we note that for r large, the second term in this last equation is negligible compared with the first. Thus,

129

$$\Phi(r, \theta) \simeq Ar\cos\theta, \qquad r \text{ large.}$$

The Gradient

We satisfy the second boundary condition by choosing $A = -E_0$. The complete solution is thus

$$\Phi(r, \theta) = -E_0 r \left(1 - \frac{R^2}{r^2} \right) \cos \theta.$$

To find the electric field, we proceed as usual with $\mathbf{E} = -\nabla\Phi$. Using Equation (IV–3), we get

$$E_r = -\frac{\partial \Phi}{\partial r} = E_0 \left[1 + \left(\frac{R}{r} \right)^2 \right] \cos \theta,$$

$$E_\theta = -\frac{1}{r} \frac{\partial \Phi}{\partial \theta} = -E_0 \left[1 - \left(\frac{R}{r} \right)^2 \right] \sin \theta,$$

$$E_z = -\frac{\partial \Phi}{\partial z} = 0.$$

You should verify that for large r, this field reduces to $E_0\mathbf{i}$ as required.

You may find this last example disquieting since a certain amount of clever guesswork is used in finding the potential. Actually there are standard procedures, which, in problems of this kind, lead more or less straightforwardly to the solution. A discussion of these procedures, however, would be very lengthy and (in the well-worn phrase) beyond the scope of this text. Before moving on, however, one further point is worth making: A solution of Laplace's equation which satisfies appropriate boundary conditions is *unique*. That is to say, there is one and only one such solution, so that if we solve a problem by guesswork and skullduggery, and someone else solves it with refined and elegant mathematical techniques, the two solutions, in spite of their disparate pedigrees, must be the same. In Problem IV–24 you will be led through a proof of this remarkable fact.

Directional Derivatives and the Gradient

We have introduced the gradient as a sort of mathematical artifice useful in discussing path-independent line integrals. We now turn to a more detailed examination of the gradient in order to describe its geometrical significance.

Before beginning our discussion, we make a few comments on Taylor series since these are needed in what follows. For a scalar function of one variable which is suitably continuous and differentiable, we have

$$f(x + \Delta x) = f(x) + \Delta x f'(x) + \tfrac{1}{2}(\Delta x)^2 f''(x) + \cdots.$$

This says that the value of the function at some point $x + \Delta x$ can be written as the sum of (usually) infinitely many terms which involve the function and its derivatives at some other point x. Among other things, this Taylor series is useful for calculation, for if the two points are close together (that is, if Δx is small), then we can truncate the series after a certain number of terms (which we hope is small) since the neglected terms, each proportional to some large power of the small number Δx, will sum to a value which is negligible.

Taylor series can also be formed for functions of several variables. Thus, for a function of two variables we have

$$f(x + \Delta x, y + \Delta y)$$

$$= f(x, y) + \Delta x \frac{\partial f}{\partial x} + \Delta y \frac{\partial f}{\partial y} + \cdots. \quad \text{(IV–13)}$$

This says that the value of the function at some point $(x + \Delta x, y + \Delta y)$ can be written as a sum of (usually) infinitely many terms which involve the function and its derivatives at some other point (x, y). We shall never need the explicit form of the remaining terms of this series [represented by the dots in Equation (IV–13)]. We should know, however, that these terms involve higher powers of the "small" numbers Δx and Δy (for example, Δx^2, Δy^2, $\Delta x \Delta y$, Δx^3, Δy^3, $\Delta x^2 \Delta y$, and so on). With these simple ideas in mind we turn now to our main task.

Consider some function $z = f(x, y)$. Geometrically this represents a surface as shown in Figure IV–8(a). Let (x, y) be the coordinates of a point P in the xy-plane. The height of the surface above this point is represented by the length of the dotted line PQ; that is, $PQ = z = f(x, y)$. Suppose now we take a short step in the xy-plane to a new point P' with coordinates $(x + \Delta x, y + \Delta y)$. The height of the surface above this point is $P'Q' = f(x + \Delta x, y + \Delta y)$. Let Δs be the length of the step $(\Delta s = PP')$.

We next ask how much the function f has changed as a result

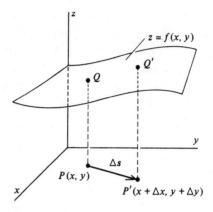

Figure IV–8(a)

of taking this step. Clearly this change is the difference in the two heights PQ and $P'Q'$, and

$$P'Q' - PQ \equiv \Delta f = f(x + \Delta x, y + \Delta y) - f(x, y).$$

Applying the Taylor series formula stated above [Equation (IV–13)], we get

$$\Delta f = f(x, y) + \Delta x \frac{\partial f}{\partial x} + \Delta y \frac{\partial f}{\partial y} + \cdots - f(x, y)$$

$$= \Delta x \frac{\partial f}{\partial x} + \Delta y \frac{\partial f}{\partial y} + \cdots.$$

We now recast this expression by what at first may seem an unnecessary elaboration of the notation. Let $\Delta \mathbf{s}$ be a vector that has magnitude Δs and points from P to P'. Clearly,

$$\Delta \mathbf{s} = \mathbf{i}\Delta x + \mathbf{j}\Delta y.$$

But the gradient of f is

$$\nabla f = \mathbf{i} \frac{\partial f}{\partial x} + \mathbf{j} \frac{\partial f}{\partial y}$$

(an obvious specialization of the gradient notation to a function of two, rather than three, variables). It follows at once that

$$\Delta f = (\Delta \mathbf{s}) \cdot (\nabla f) + \cdots.$$

Complicating matters slightly more, let $\hat{\mathbf{u}}$ be a unit vector in the direction of $\Delta\mathbf{s}$. Then

$$\Delta\mathbf{s} = \hat{\mathbf{u}}\Delta s$$

and

$$\Delta f = (\hat{\mathbf{u}} \cdot \nabla f)\,\Delta s + \cdots,$$

so that

$$\frac{\Delta f}{\Delta s} = \hat{\mathbf{u}} \cdot \nabla f + \cdots.$$

We now take the limit of this equation to get

$$\frac{df}{ds} \equiv \lim_{\Delta s \to 0} \frac{\Delta f}{\Delta s} = \hat{\mathbf{u}} \cdot \nabla f. \qquad \text{(IV–14)}$$

There is no longer any need of "$+ \cdots$" since the dots represented terms which go to zero as Δs goes to zero.

This new expression [Equation (IV–14)] has a simple interpretation: it is the *rate* of change of the function $f(x, y)$ in the direction of $\Delta\mathbf{s}$ (that is, of $\hat{\mathbf{u}}$). Redrawing Figure IV–8(a) and passing a plane through P and P' parallel to the z-axis [Figure IV–8(b)], we see that it cuts the surface $z = f(x, y)$ in a curve C.

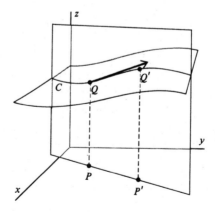

Figure IV–8(b)

The quantity df/ds defined in Equation (IV–14) is the slope of this curve at the point Q.

The quantity *df/ds* is called the _directional derivative of f._ Although the analysis given above which led to this derivative was for functions of two variables, the results all apply to functions of three (or more) variables. Thus,

$$\frac{d}{ds} F(x, y, z) = \hat{\mathbf{u}} \cdot \nabla F$$

is the rate of change of the function $F(x, y, z)$ in the direction specified by the unit vector $\hat{\mathbf{u}}$.

An example of the directional derivative may be amusing here. We'll work with a function of two variables so that we can draw pictures. Thus, let's consider

$$z = f(x, y) = (x^2 + y^2)^{1/2},$$

which is an inverted right circular cone whose axis coincides with the z-axis [see Figure IV–9(a)]. We ask for the directional deriv-

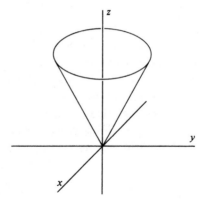

Figure IV–9(a)

ative of this function at some point $x = a$ and $y = b$ and in the direction specified by $\hat{\mathbf{u}} = \mathbf{i} \cos \theta + \mathbf{j} \sin \theta$ [see Figure IV–9(b)].

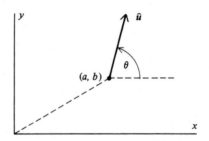

Figure IV–9(b)

First we need the gradient of $f(x, y)$. But

$$\frac{\partial f}{\partial x} = \frac{x}{z} \quad \text{and} \quad \frac{\partial f}{\partial y} = \frac{y}{z},$$

as you can easily verify. Thus,

$$\nabla f = \frac{\mathbf{i}x + \mathbf{j}y}{z}$$

and

$$\frac{df}{ds} = \hat{\mathbf{u}} \cdot \nabla f = \frac{x \cos \theta + y \sin \theta}{z} \rightarrow \frac{a \cos \theta + b \sin \theta}{\sqrt{a^2 + b^2}}.$$

Suppose θ is chosen so that $\hat{\mathbf{u}}$ is in the radial direction as indicated in Figure IV–9(c). This means

$$\cos \theta = \frac{a}{(a^2 + b^2)^{1/2}},$$

$$\sin \theta = \frac{b}{(a^2 + b^2)^{1/2}},$$

and so

$$\frac{df}{ds} = \frac{a}{\sqrt{a^2 + b^2}} \cdot \frac{a}{\sqrt{a^2 + b^2}} + \frac{b}{\sqrt{a^2 + b^2}} \cdot \frac{b}{\sqrt{a^2 + b^2}} = 1.$$

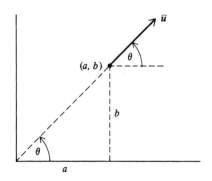

Figure IV–9(c)

The significance of this result is brought out in Figure IV–9(d).

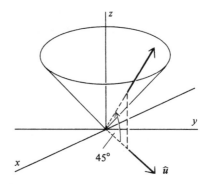

Figure IV–9(d)

A second interesting case is that in which **û** is chosen perpendicular to the direction of the previous example [see Figure IV–9(e)]. We then have

$$\cos \theta = \frac{-b}{(a^2 + b^2)^{1/2}},$$

$$\sin \theta = \frac{a}{(a^2 + b^2)^{1/2}},$$

and so

$$\frac{df}{ds} = \frac{a}{\sqrt{a^2 + b^2}} \left(-\frac{b}{\sqrt{a^2 + b^2}} \right)$$

$$+ \frac{b}{\sqrt{a^2 + b^2}} \left(\frac{a}{\sqrt{a^2 + b^2}} \right) = 0.$$

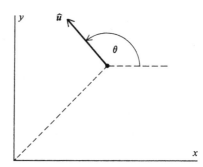

Figure IV–9(e)

The meaning of this result is illustrated in Figure IV–9(f).

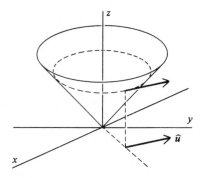

Figure IV–9(f)

Geometric Significance of the Gradient

With the concept of the directional derivative at our disposal, we are now in a position to give a geometric interpretation of the gradient. At some point P_0 with coordinates (x_0, y_0, z_0) we have

$$\left(\frac{dF}{ds}\right)_0 = \hat{\mathbf{u}} \cdot (\nabla F)_0,$$

where the subscript "0" means the quantity is to be evaluated at the point (x_0, y_0, z_0). Now $(\nabla F)_0$, the gradient of F evaluated at P_0, may be represented by an arrow emanating from that point as shown in Figure IV–10. If we ask in what direction we must move to make $(dF/ds)_0$ as large as possible, it is clear that $\hat{\mathbf{u}}$

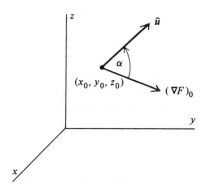

137   Figure IV–10

should be in the same direction as $(\nabla F)_0$. This is because if we let α be the angle between $\hat{u}$ and $(\nabla F)_0$, then $(dF/ds)_0 = |\nabla F|_0 \cos \alpha$, and this is as large as it can be when $\alpha = 0$. Thus, *the gradient of a scalar function F(x, y, z) is a vector that is in the direction in which F undergoes the greatest rate of increase and that has magnitude equal to the rate of increase in that direction.*

To illustrate this interpretation of the gradient, let us go back to the inverted cone $z = f(x, y) = (x^2 + y^2)^{1/2}$ we discussed above. We learned that

$$\nabla f = \frac{\mathbf{i}x + \mathbf{j}y}{z}$$

and

$$\frac{df}{ds} = \frac{a \cos \theta + b \sin \theta}{\sqrt{a^2 + b^2}} \equiv D(\theta).$$

To find the direction in which $f(x, y)$ undergoes the greatest rate of change, we set

$$\frac{dD}{d\theta} = \frac{-a \sin \theta + b \cos \theta}{\sqrt{a^2 + b^2}} = 0.$$

This gives $\tan \theta = b/a$, whence $\cos \theta = a/(a^2 + b^2)^{1/2}$ and $\sin \theta = b/(a^2 + b^2)^{1/2}$. So $(df/ds)_{max} = 1$. On the other hand,

$$|\nabla f| = \left[\frac{x^2 + y^2}{z^2} \right]^{1/2} = 1,$$

since $z^2 = x^2 + y^2$. Furthermore, $\tan \theta = b/a$ corresponds to the direction $a\mathbf{i} + b\mathbf{j}$, while at the point (a, b),

$$\nabla f = \frac{a\mathbf{i} + b\mathbf{j}}{(a^2 + b^2)^{1/2}},$$

which is a vector in the same direction. Thus both properties of the gradient are illustrated; it's in the direction of maximum rate of increase, and its magnitude is equal to the rate of increase in that direction.

With this geometric interpretation of the gradient at our disposal, we can now see the reason for the negative sign in the equation $\mathbf{E} = -\nabla \Phi$: Since $\nabla \Phi$ is a vector in the direction of increasing Φ, the force on a *positive* charge q is

$\mathbf{F} = q\mathbf{E} = -q\nabla\Phi$, which is in the direction of *decreasing* Φ. Thus, the negative sign ensures that a positive charge moves "downhill" from a higher to a lower potential.

There is another property of the gradient useful in understanding its geometric significance. To make this discussion concrete, let $T(x, y, z)$ be a scalar function which gives the temperature at any point (x, y, z). The locus of all points having the same temperature T_0 is (in the simplest case) a surface whose equation is $T(x, y, z) = T_0$ (Figure IV–11). This is called an isothermal surface. We now show that ∇T is a vector normal to the isothermal

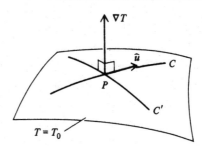

Figure IV–11

surface. Let C be any curve lying in the isothermal surface and let P be any point on C. Let $\hat{\mathbf{u}}$ be the unit vector tangent to C at P (it doesn't matter which direction along C we take). The directional derivative in the direction $\hat{\mathbf{u}}$ is

$$\left(\frac{dT}{ds}\right) = \hat{\mathbf{u}} \cdot \nabla T = 0$$

because T does not change as we move along the isothermal surface. If the scalar product of two vectors, neither of them zero, vanishes, the two vectors are perpendicular. Thus ∇T is perpendicular to C at P. By the same argument it is perpendicular to *any* curve on the surface through P (such as C' in Figure IV–11). But this can be true only if ∇T is normal to the isothermal surface at P. In general then, $\nabla f(x, y, z)$, *where* f(x, y, z) *is a scalar function, is normal to the surface* f(x, y, z) = *constant.*[3]

A simple example of this property of the gradient is provided by the function $F(x, y, z) = x^2 + y^2 + z^2$. The surface $F(x, y, z)$ = constant is, of course, a sphere (assuming the constant is pos-

---

[3] The connection between this property of the gradient and our earlier expression for the unit vector normal to a surface [Equation (II–4)] is the subject of Problem IV–20.

itive). As you should verify for yourself, $\nabla F = 2(\mathbf{i}x + \mathbf{j}y + \mathbf{k}z)$ $= 2\mathbf{r}$. Thus, we have a familiar result: A vector normal to a spherical surface is in the radial direction. We'll leave it to you to ponder the geometric relation between the electrostatic field **E** and its *equipotential surfaces* $\Phi(x, y, z)$ = constant.

We can make a simple connection between the property of the gradient just discussed and the fact that it is in the direction of the greatest rate of increase. Any displacement from the surface $f(x, y, z)$ = constant, regarded as a vector **s**, can be resolved into a component along the surface $(s_\parallel)$ and one normal to it $(s_\perp)$, as shown in Figure IV–12. That part of the displacement along the

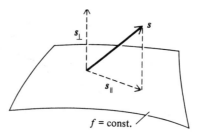

Figure IV–12

surface is "wasted motion" if our aim in moving is to cause a change in the value of $f(x, y, z)$. Only the normal component carries us away from the surface and causes a change in f. From this it is clear that the greatest increase possible for a given magnitude of displacement should occur when we move away from the surface in the normal direction. But we have already established that the greatest rate of increase occurs in the direction of the gradient. Thus the gradient is normal to the surface.

The Gradient in Cylindrical and Spherical Coordinates

A by-product of our discussion of the directional derivative is the "easier and faster" method for calculating the gradient in spherical and cylindrical coordinates mentioned earlier (see page 120). To determine this method, we begin by outlining our derivation of df/ds:[4]

[4] The calculation outlined here pertains to a function of three variables and is a simple generalization of the calculation on pages 130–33 which deals with a function of two variables.

1. Our first step is to consider a scalar function of three Cartesian coordinates $f(x, y, z)$ and use Taylor series to determine the change in f caused by a displacement from the point (x, y, z) to a second point $(x + \Delta x, y + \Delta y, z + \Delta z)$. We find for this change

$$\Delta f = \frac{\partial f}{\partial x} \Delta x + \frac{\partial f}{\partial y} \Delta y + \frac{\partial f}{\partial z} \Delta z + \cdots.$$

2. We next write Δf in terms of $\Delta \mathbf{s}$, the vector displacement from (x, y, z) to $(x + \Delta x, y + \Delta y, z + \Delta z)$. Clearly (see Figure IV–13)

$$\Delta \mathbf{s} = \mathbf{i}\Delta x + \mathbf{j}\Delta y + \mathbf{k}\Delta z,$$

so that

$$\Delta f = \left(\mathbf{i} \frac{\partial f}{\partial x} + \mathbf{j} \frac{\partial f}{\partial y} + \mathbf{k} \frac{\partial f}{\partial z} \right) \cdot \Delta \mathbf{s} + \cdots.$$

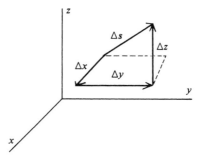

Figure IV–13

3. Finally we write $\Delta \mathbf{s} = \hat{\mathbf{u}}\Delta s$, divide by Δs, and take the limit:

$$\lim_{\Delta s \to 0} \frac{\Delta f}{\Delta s} \equiv \frac{df}{ds} = \left(\mathbf{i} \frac{\partial f}{\partial x} + \mathbf{j} \frac{\partial f}{\partial y} + \mathbf{k} \frac{\partial f}{\partial z} \right) \cdot \hat{\mathbf{u}}.$$

The quantity which is dotted into $\hat{\mathbf{u}}$ in this last expression is then recognized as the gradient of f in Cartesian coordinates.

To obtain the gradient of a scalar function in cylindrical coordinates we proceed in much the same way:

1. We consider a scalar function of three cylindrical coordinates, $f(r, \theta, z)$. Using Taylor series, we find the change in f due to a displacement from the point (r, θ, z) to a second point $(r + \Delta r, \theta + \Delta\theta, z + \Delta z)$:

$$\Delta f = \frac{\partial f}{\partial r} \Delta r + \frac{\partial f}{\partial \theta} \Delta\theta + \frac{\partial f}{\partial z} \Delta z + \cdots.$$

2. Next, we write Δf in terms of $\Delta\mathbf{s}$. This is the heart of the calculation. From Figure IV–14 we have

$$\Delta\mathbf{s} = \hat{\mathbf{e}}_r \Delta r + \hat{\mathbf{e}}_\theta r \Delta\theta + \hat{\mathbf{e}}_z \Delta z.$$

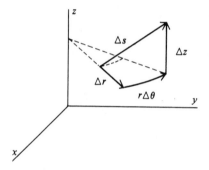

Figure IV–14

There are two features of this expression which require some discussion. First, the displacement in the direction of increasing θ (of magnitude $r \Delta\theta$) is an arc of a circle rather than a straight line segment. However, since we will eventually pass to the limit as $\Delta s \to 0$, we may regard $\Delta\theta$ (as well as Δr and Δz) as arbitrarily small, in which case the arc is arbitrarily close to its subtending chord. Thus, as indicated in Figure IV–15, Δr, $r \Delta\theta$, and Δz approximate to any desired degree of accuracy three mutually perpendicular displacements, the analogs of the three Cartesian displacements Δx, Δy, and Δz (see Figure IV–13).

The second feature of our expression for $\Delta\mathbf{s}$ which requires comment also has to do with the displacement in the direction of increasing θ. It is this: Since the arc is part

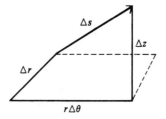

Figure IV–15

of a circle of radius $r + \Delta r$, we should, strictly speaking, write the displacement as $(r + \Delta r) \Delta\theta$, not $r \Delta\theta$. But the additional term $\Delta r \Delta\theta$ is "second order"; that is, it is the product of two small quantities and therefore negligible compared with $r \Delta\theta$.

If we now write our expression for Δf in terms of Δs, we get

$$\Delta f = \left(\hat{e}_r \frac{\partial f}{\partial r} + \hat{e}_\theta \frac{1}{r} \frac{\partial f}{\partial \theta} + \hat{e}_z \frac{\partial f}{\partial z} \right) \cdot \Delta s + \cdots.$$

Note the factor $1/r$ in the second term to compensate for the factor r in $\hat{e}_\theta r \Delta\theta$ in Δs.

3. Finally, putting $\Delta s = \hat{u} \Delta s$, we find

$$\lim_{\Delta s \to 0} \frac{\Delta f}{\Delta s} \equiv \frac{df}{ds} = \left(\hat{e}_r \frac{\partial f}{\partial r} + \hat{e}_\theta \frac{1}{r} \frac{\partial f}{\partial \theta} + \hat{e}_z \frac{\partial f}{\partial z} \right) \cdot \hat{u}.$$

The quantity in the above expression dotted into $\hat{u}$ is the gradient of f in cylindrical coordinates.

An analogous procedure can be used to find the gradient in spherical coordinates; this has been left as an exercise (see Problem IV–22).

PROBLEMS

IV-1   (a) Calculate $\mathbf{F} = \nabla f$ for each of the following scalar functions:
   (i) $f = xyz$.
   (ii) $f = x^2 + y^2 + z^2$.
   (iii) $f = xy + yz + xz$.
   (iv) $f = 3x^2 - 4z^2$.
   (v) $f = e^{-x} \sin y$.

143

(b) Verify that

$$\oint_C \mathbf{F} \cdot \hat{\mathbf{t}} \, ds = 0$$

for one or more of the functions **F** determined in part (a) choosing for the curve C:

    (i) the square in the xy-plane with vertices at $(0, 0)$, $(1, 0)$, $(1, 1)$, and $(0, 1)$.

    (ii) the triangle in the yz-plane with vertices at $(0, 0)$, $(1, 0)$, and $(0, 1)$.

    (iii) the circle of unit radius centered at the origin and lying in the xz-plane.

(c) Verify by direct calculation that $\nabla \times \mathbf{F} = 0$ for one or more of the functions **F** determined in part (a).

IV–2 Verify the following identities in which f and g are arbitrary differentiable scalar functions of position, and **F** and **G** are arbitrary differentiable vector functions of position.

    (a) $\nabla(fg) = f\nabla g + g\nabla f$.

    (b) $\nabla(\mathbf{F} \cdot \mathbf{G}) = (\mathbf{G} \cdot \nabla)\mathbf{F} + (\mathbf{F} \cdot \nabla)\mathbf{G} + \mathbf{F} \times (\nabla \times \mathbf{G}) + \mathbf{G} \times (\nabla \times \mathbf{F})$.

    (c) $\nabla \cdot (f\mathbf{F}) = f\nabla \cdot \mathbf{F} + \mathbf{F} \cdot \nabla f$.

    (d) $\nabla \cdot (\mathbf{F} \times \mathbf{G}) = \mathbf{G} \cdot (\nabla \times \mathbf{F}) - \mathbf{F} \cdot (\nabla \times \mathbf{G})$.

    (e) $\nabla \times (f\mathbf{F}) = f\nabla \times \mathbf{F} + (\nabla f) \times \mathbf{F}$.

    (f) $\nabla \times (\mathbf{F} \times \mathbf{G}) = (\mathbf{G} \cdot \nabla)\mathbf{F} - (\mathbf{F} \cdot \nabla)\mathbf{G} + \mathbf{F}(\nabla \cdot \mathbf{G}) - \mathbf{G}(\nabla \cdot \mathbf{F})$.

    (g) $\nabla \times (\nabla \times \mathbf{F}) = \nabla(\nabla \cdot \mathbf{F}) - \nabla^2\mathbf{F}$.

IV–3 Show that $\nabla \times \nabla f = 0$ where $f(x, y, z)$ is an arbitrary differentiable scalar function. Assume that mixed second-order partial derivatives are independent of the order of differentiation. For example, $\partial^2 f/\partial x \, \partial z = \partial^2 f/\partial z \, \partial x$.

IV–4 (a) Each of the following functions is smooth in a simply connected region. Determine which of them may be written as the gradient of a scalar function, and for those which can, use Equation (IV–2) to find that scalar function.

    (i) $\mathbf{F} = y\mathbf{i}$.

    (ii) $\mathbf{F} = C\mathbf{k}$, C a constant.

    (iii) $\mathbf{F} = \mathbf{i}yz + \mathbf{j}xz + \mathbf{k}xy$.

    (iv) $\mathbf{F} = \mathbf{i}x + \mathbf{j}y + \mathbf{k}z$.

    (v) $\mathbf{F} = \mathbf{i}e^{-z} \sin y + \mathbf{j}e^{-y} \sin z + \mathbf{k}e^{-x} \sin y$.

(b) Neither of the following functions is smooth everywhere. Nonetheless each can be written as the gradient of a scalar function. Use Equation (IV–2) to find that scalar function.

    (i) $\mathbf{F} = \mathbf{r}/r^2$, $\mathbf{r} = \mathbf{i}x + \mathbf{j}y$.

    (ii) $\mathbf{F} = \mathbf{r}/r^{1/2}$, $\mathbf{r} = \mathbf{i}x + \mathbf{j}y + \mathbf{k}z$.

IV–5 The function $\mathbf{F}(r, \theta, z)$ defined in Problem III–17 is smooth and

$$c \, r^{-2} ?$$

$$f = \left(x^2 + y^2 \right)^{\frac{1}{2}} \quad {}_z = f = h^{\frac{1}{2}}$$

$$f = h(z)^{\frac{1}{2}}$$

$$\frac{1}{2} h(z)^{-\frac{1}{2}} \cdot \frac{dz}{dx}$$

$$\frac{1}{2\sqrt{z}} \cdot \overset{\shortmid\shortmid}{2x}$$

$$\frac{x}{\sqrt{z}}$$

Join Your Grove Ave. neighbors

for coffee and treats on

Memorial Day, Monday, May 30

9:00 a.m.—??

416 N. Grove

Bring a Treat to share and it will be gladly eaten!

hosts: jack & john barclay, mark, janelle, chris, & dave lobmier has zero curl in a nonsimply connected region consisting of all of three-dimensional space with the z-axis removed. Show that there is no scalar function ψ such that $\mathbf{F} = \nabla\psi$ by evaluating the line integral of $\mathbf{F} \cdot \hat{\mathbf{t}}$ from the point $P_1(0, -1, 0)$ to the point $P_2(0, 1, 0)$ over two different paths: C_R, the right-hand side of the circle of radius 1 lying in the xy-plane and centered at the origin (see figure), and C_L, the left-hand side of the same circle. Orient the paths as shown. Why does the fact that the two paths give different results imply that there is no scalar function ψ such that $\mathbf{F} = \nabla\psi$?

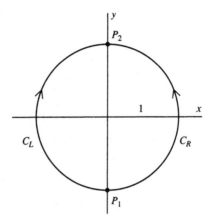

IV–6   (a) An electric dipole of strength p situated at the origin and oriented in the positive z-direction gives rise to an electrostatic field

$$\mathbf{E}(r, \theta, \phi) = \frac{1}{4\pi\epsilon_0} \frac{p}{r^3} (2\hat{\mathbf{e}}_r \cos\theta + \hat{\mathbf{e}}_\theta \sin\theta).$$

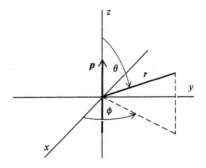

Use Equation (IV–2) to show that the dipole potential is given by

$$\Phi(r, \theta, \phi) = \frac{1}{4\pi\epsilon_0} \frac{p \cos\theta}{r^2}.$$

Useful information: In spherical coordinates,

$$\hat{\mathbf{t}} = \hat{\mathbf{e}}_r \frac{dr}{ds} + \hat{\mathbf{e}}_\theta r \frac{d\theta}{ds} + \hat{\mathbf{e}}_\phi r \sin\theta \frac{d\phi}{ds} .$$

(b) Calculate the flux of the dipole field through a sphere of radius R centered at the origin.

(c) What is the flux of the dipole field over *any* closed surface which does not pass through the origin?

IV–7 Here is a "proof" that there is no such thing as magnetism. One of Maxwell's equations tells us that

$$\nabla \cdot \mathbf{B} = 0,$$

where $\mathbf{B}$ is any magnetic field. Then using the divergence theorem, we find

$$\iint_S \mathbf{B} \cdot \hat{\mathbf{n}} \, dS = \iiint_V \nabla \cdot \mathbf{B} \, dV = 0.$$

Because $\mathbf{B}$ has zero divergence, we know (see Problem III–24) there exists a vector function, call it $\mathbf{A}$, such that

$$\mathbf{B} = \nabla \times \mathbf{A}.$$

Combining these last two equations, we get

$$\iint_S \hat{\mathbf{n}} \cdot \nabla \times \mathbf{A} \, dS = 0.$$

Next we apply Stokes' theorem and the above result to find

$$\oint_C \mathbf{A} \cdot \hat{\mathbf{t}} \, ds = \iint_S \hat{\mathbf{n}} \cdot \nabla \times \mathbf{A} \, dS = 0.$$

Thus we have shown that the circulation of $\mathbf{A}$ is path-independent. It follows that we can write $\mathbf{A} = \nabla\psi$ where ψ is some scalar function. Since the curl of the gradient of a function is zero, we arrive at the remarkable fact that

$$\mathbf{B} = \nabla \times \nabla\psi = 0;$$

that is, all magnetic fields are zero! Where did we go wrong? [Taken from G. Arfken, *Amer. J. Phys.*, **27**, 526 (1959).]

IV–8 Fick's law states that in certain diffusion processes the current density $\mathbf{J}$ is proportional to the negative of the gradient of the density ρ; that is, $\mathbf{J} = -k\nabla\rho$, where k is a positive constant. If a substance of density $\rho(x, y, z, t)$ and velocity $\mathbf{v}(x, y, z, t)$ diffuses according to Fick's law, show that the flow is *irrotational* (that is, $\nabla \times \mathbf{v} = 0$).

IV–9 (a) A substance diffuses according to Fick's law (see Problem IV–8). Assuming the diffusing matter is conserved, derive the diffusion equation

$$\frac{\partial \rho}{\partial t} = k \nabla^2 \rho.$$

(b) Bacteria of density ρ diffuse in a medium according to Fick's law and reproduce at a rate $\lambda \rho$ per unit volume (λ is a positive constant). Show that

$$\frac{\partial \rho}{\partial t} = k \nabla^2 \rho + \lambda \rho.$$

IV–10 (a) A fluid is said to be *incompressible* if its density ρ is a constant (that is, is independent of x, y, z, and t). Use the continuity equation to show that the velocity $\mathbf{v}$ of an incompressible fluid satisfies the equation $\nabla \cdot \mathbf{v} = 0$.

(b) If $\nabla \times \mathbf{v} = 0$, the fluid flow is said to be *irrotational*. Show that for an incompressible fluid undergoing irrotational flow,

$$\nabla^2 \phi = 0,$$

where ϕ, a scalar function called the *velocity potential*, is so defined that $\mathbf{v} = \nabla \phi$.

IV–11 The heat Q in a body of volume V is given by

$$Q = c \iiint_V T \rho \, dV,$$

where c is a constant called the specific heat of the body, and $T(x, y, z, t)$ and $\rho(x, y, z)$ are, respectively, the temperature and density of the body. (Note that we are assuming the density to be independent of time.) The rate at which heat flows through S, the bounding surface of the body, is given by

$$\frac{dQ}{dt} = k \iint_S \hat{\mathbf{n}} \cdot \nabla T \, dS,$$

where k (assumed constant) is the thermal conductivity of the body, and the integral is taken over the surface S bounding the body. Use these facts to derive the heat flow equation

$$\nabla^2 T = \alpha \frac{\partial T}{\partial t},$$

where $\alpha = c\rho/k$.

IV–12 In nonrelativistic quantum mechanics a particle of mass m moving in a potential $V(x, y, z)$ is described by the Schrödinger equation

$$-\frac{\hbar^2}{2m} \nabla^2 \psi + V\psi = i\hbar \frac{\partial \psi}{\partial t},$$

147

where $\hbar$ is Planck's constant divided by 2π and $\psi(x, y, z, t)$, which is complex, is called the wave function. The quantity $\rho = \psi^*\psi$ is interpreted as the probability density.

(a) Use the Schrödinger equation to derive an equation of the form

$$\frac{\partial \rho}{\partial t} + \nabla \cdot \mathbf{J} = 0$$

and obtain thereby an expression for $\mathbf{J}$ in terms of ψ, ψ^*, m, and $\hbar$.

(b) Give an interpretation of $\mathbf{J}$ and of the equation derived in (a).

IV–13 (a) Find the charge density $\rho(x, y, z)$ which produces the electric field

$$\mathbf{E} = g(\mathbf{i}x + \mathbf{j}y + \mathbf{k}z),$$

where g is a constant.

(b) Find an electrostatic potential Φ such that $-\nabla\Phi$ is the field $\mathbf{E}$ given in (a).

(c) Verify that $\nabla^2\Phi = -\rho/\epsilon_0$.

IV–14 (a) Starting with the divergence theorem, derive the equation

$$\iint_S \hat{\mathbf{n}} \cdot (u\nabla v) \, dS = \iiint_V [u\nabla^2 v + (\nabla u) \cdot (\nabla v)] \, dV,$$

where u and v are scalar functions of position and S is a closed surface enclosing the volume V. This is sometimes called the first form of Green's theorem.

(b) If $\nabla^2 u = 0$ use the first form of Green's theorem to show that

$$\iint_S \hat{\mathbf{n}} \cdot (u\nabla u) \, dS = \iiint_V |\nabla u|^2 \, dV,$$

where $|\nabla u|^2 = (\nabla u) \cdot (\nabla u)$.

(c) Use the first form of Green's theorem to show that

$$\iint_S \hat{\mathbf{n}} \cdot (u\nabla v - v\nabla u) \, dS = \iiint_V (u\nabla^2 v - v\nabla^2 u) \, dV.$$

This is the second form of Green's theorem.

IV–15 An equation of the form

$$\nabla^2 f = \frac{1}{v^2} \frac{\partial^2 f}{\partial t^2},$$

where f is a twice-differentiable function of position and time, is called a wave equation. It describes a wave propagating in space with velocity v. Use Maxwell's equations (Problem III–20) to show that in the absence of charges and currents (that is, ρ and $\mathbf{J}$ both zero), all three

Cartesian components of both **E** and **B** satisfy a wave equation with $v = c$, where $c = 1/\sqrt{\epsilon_0\mu_0}$ is the velocity of light. For example,

$$\nabla^2 E_x = \frac{1}{c^2}\frac{\partial^2 E_x}{\partial t^2}.$$

Thus, the existence of electromagnetic waves traveling in empty space with the velocity of light is a consequence of Maxwell's equations.

IV–16 (a) In the text we found the potential and field for the case of an infinite cylinder between parallel plates with the cylinder held at zero potential. How must the solution be modified if the cylinder is held at a potential $V_0 \neq 0$?

(b) Show that there is no net charge on the cylinder.

IV–17 (a) A sphere of radius R is situated between two very large parallel plates which are separated by a distance s. A potential difference is maintained between the plates and the sphere is held at zero potential. Find the potential and field everywhere outside the sphere and between the plates. Assume that $R \ll s$.

(b) Show that there is no net charge on the sphere.

(c) Repeat part (a) assuming the sphere is held at a potential $V_0 \neq 0$.

IV–18 Let $f(x, y)$ be a differentiable scalar function of x and y, and let $\hat{u} = \mathbf{i}\cos\theta + \mathbf{j}\sin\theta$. Transform to a rotated coordinate system x', y' such that x' is parallel to $\hat{u}$ (see the figure). Show that the directional derivative in the direction of $\hat{u}$

$$\frac{df}{ds} = \hat{u}\cdot\nabla f = \frac{\partial f}{\partial x'}.$$

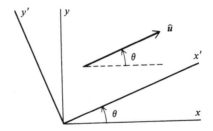

IV–19 You are at a point (a, b, c) on the surface

$$z = (r^2 - x^2 - y^2)^{1/2} \qquad (z \geq 0).$$

Assuming both a and b are positive, in what direction must you move

(a) so that the rate of change of z will be zero?

(b) so that the rate of increase of z will be greatest?

(c) so that the rate of decrease of z will be greatest?

Draw a sketch to show the geometric significance of your answers.

IV–20 The unit vector normal to the surface $z = f(x, y)$ is given by

$$\hat{\mathbf{n}} = \left(-\mathbf{i}\frac{\partial f}{\partial x} - \mathbf{j}\frac{\partial f}{\partial y} + \mathbf{k} \right) \Big/ \sqrt{1 + \left(\frac{\partial f}{\partial x}\right)^2 + \left(\frac{\partial f}{\partial y}\right)^2}$$

[see Equation (II–4)]. We have also established that ∇F is a vector normal to the surface $F(x, y, z) = $ const. (page 139) so that $\nabla F/|\nabla F|$ is a unit vector normal to the surface $F(x, y, z) = $ const. Show that these two expressions for the unit normal vector are identical if $F(x, y, z) = $ const. and $z = f(x, y)$ describe the same surface.

IV–21 Use the results of Problem II–18 and the expression for the gradient in cylindrical coordinates (see page 143) to obtain the form of the Laplacian in cylindrical coordinates given on page 128.

IV–22 Using the procedure outlined in the text (pages 142–43) obtain the expression for the gradient of ψ in spherical coordinates:

$$\nabla\psi = \hat{\mathbf{e}}_r \frac{\partial\psi}{\partial r} + \hat{\mathbf{e}}_\theta \frac{1}{r}\frac{\partial\psi}{\partial\theta} + \hat{\mathbf{e}}_\phi \frac{1}{r\sin\theta}\frac{\partial\psi}{\partial\phi}.$$

IV–23 Use the results of Problem II–19 and the expression for the gradient in spherical coordinates derived in Problem IV–22 to obtain the form of the Laplacian in spherical coordinates given on page 125.

IV–24 Suppose you find a solution of Laplace's equation which satisfies certain boundary conditions. Is this solution unique or are there others? This problem will answer that question in certain simple cases. Consider the region of space completely enclosed by a surface S_0 and containing in its interior objects 1, 2, 3, . . . (two of which are pictured in the diagram). Suppose that S_0 is maintained at a constant potential Φ_0, object no. 1 at Φ_1, object no. 2 at Φ_2, and so on. Then in the charge-free region R enclosed by S_0 and between the objects, the potential must satisfy Laplace's equation

$$\nabla^2\Phi = 0$$

and the boundary conditions

$$\Phi = \begin{cases} \Phi_0 \text{ on } S_0 \\ \Phi_1 \text{ on } S_1 \\ \Phi_2 \text{ on } S_2 \\ \vdots \end{cases}$$

The following steps will guide you through a proof that Φ is unique.

(a) Assume that there are two potentials u and v, both of which satisfy Laplace's equation and the boundary conditions listed above. Form their difference $w = u - v$. Show that $\nabla^2 w = 0$ in R.

(b) What are the boundary conditions satisfied by w?

(c) Apply the divergence theorem to

$$\iint_S \hat{\mathbf{n}} \cdot (w \nabla w) \, dS,$$

where the integration is carried out over the surface $S_0 + S_1 + S_2 + \cdots$, and show thereby that

$$\iiint_V |\nabla w|^2 \, dV = 0.$$

where V is the volume of the region R.

(d) From the result of (c) argue that $\nabla w = 0$ and that this, in turn, means w is a constant.

(e) If w is a constant, what is its value? (Use the boundary conditions on w to answer this.) What does this say about u and v?

(f) The uniqueness proof outlined in (a) to (e) involves specifying the value of the potential on various surfaces. Might we have specified a different kind of boundary condition and still proved uniqueness? If so, in what way or ways would the proof and the result differ from those given above?

IV–25 In the text we defined the gradient in terms of certain partial derivatives. It is possible to give an alternative definition similar in form to our definitions of the divergence and the curl. Thus,

$$\nabla f = \lim_{\Delta V \to 0} \frac{1}{\Delta V} \iint_S \hat{\mathbf{n}} f \, dS.$$

Here f is a scalar function of position, S a closed surface, and ΔV the volume it encloses. As usual, $\hat{\mathbf{n}}$ is a unit vector normal to S and pointing out from the enclosed volume.

(a) Following a procedure similar to the one used in the text in treating the divergence, integrate over a "cuboid" and show that the definition given above yields the expression

$$\nabla f = \mathbf{i} \frac{\partial f}{\partial x} + \mathbf{j} \frac{\partial f}{\partial y} + \mathbf{k} \frac{\partial f}{\partial z}.$$

(b) Use the alternative definition of the gradient given above to show that the directional derivative of f in the direction specified by the unit vector $\hat{\mathbf{u}}$ is given by

$$\frac{df}{ds} = \hat{\mathbf{u}} \cdot \nabla f.$$

[*Hint:* Evaluate

$$\hat{u} \cdot \iint_S \hat{n} f \, dS = \iint_S \hat{u} \cdot \hat{n} f \, dS$$

over a small cylinder (length Δs, cross-sectional area ΔA; see figure) whose axis is in the direction of the constant unit vector $\hat{u}$. Then divide by the volume of the cylinder ($\Delta s \, \Delta A$) and take the limit as the volume approaches zero.]

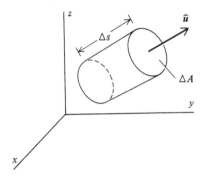

(c) Arguing as we did in the text in establishing the divergence theorem, use the alternative definition of the gradient to show that

$$\iint_S \hat{n} f \, dS = \iiint_V \nabla f \, dV,$$

where S is a closed surface enclosing the volume V.

(d) Obtain the relation stated in (c) directly from the divergence theorem. [*Hint:* In $\iint_S \mathbf{F} \cdot \hat{n} \, dS = \iiint_V \nabla \cdot \mathbf{F} \, dV$ put $\mathbf{F} = \hat{e} f$ where $\hat{e}$ is a constant unit vector.]

(e) Verify the relation stated in (c) for the scalar function

$$f = x^2 + y^2 + z^2$$

integrating over the unit cylinder shown in the figure.

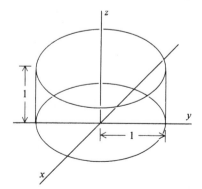

IV–26 (a) Consider a surface $z = f(x, y)$. Let **u** be a vector of arbitrary length tangent to the surface at a point $P(x, y, z)$ in the direction of the unit vector $\hat{\mathbf{p}} = i p_x + j p_y$ as indicated in the figure. Use the directional derivative to show that

$$\mathbf{u} = \hat{\mathbf{p}} + \mathbf{k}(\hat{\mathbf{p}} \cdot \nabla f),$$

where ∇f is evaluated at (x, y). [*Note:* Since the length of **u** is arbitrary, your result may differ from the above by some positive multiplicative constant.]

(b) Let **v** be a second vector of arbitrary length tangent to the surface at P but in the direction of the unit vector $\hat{\mathbf{q}} = i q_x + j q_y$ ($\hat{\mathbf{p}} \neq \hat{\mathbf{q}}$). Then from (a) we have

$$\mathbf{v} = \hat{\mathbf{q}} + \mathbf{k}(\hat{\mathbf{q}} \cdot \nabla f).$$

Show that

$$\mathbf{u} \times \mathbf{v} = [\mathbf{k} \cdot (\hat{\mathbf{p}} \times \hat{\mathbf{q}})](\mathbf{k} - \nabla f)$$

and use this to rederive Equation (II–4) for the unit vector $\hat{\mathbf{n}}$ normal to the surface $z = f(x, y)$ at (x, y, z). This shows that the result derived in the text for $\hat{\mathbf{n}}$ is unique (apart from sign) even though it was obtained with the special choices $\hat{\mathbf{p}} = \mathbf{i}$ and $\hat{\mathbf{q}} = \mathbf{j}$.

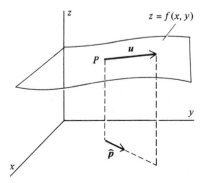

IV–27 (a) Using Maxwell's equations (see Problem III–20), show that we can write

$$\mathbf{B} = \nabla \times \mathbf{A},$$

$$\mathbf{E} = -\nabla \Phi - \frac{\partial \mathbf{A}}{\partial t},$$

where **A** (called the vector potential) is some vector function of position and time, and Φ (the scalar potential) is some scalar function of position and time, provided **A** and Φ satisfy the equations

$$\nabla^2\Phi + \frac{\partial}{\partial t}(\nabla \cdot \mathbf{A}) = -\rho/\epsilon_0,$$

$$\nabla^2\mathbf{A} - \mu_0\epsilon_0 \frac{\partial^2\mathbf{A}}{\partial t^2} = -\mu_0\mathbf{J} + \nabla\left[\nabla \cdot \mathbf{A} + \mu_0\epsilon_0 \frac{\partial\Phi}{\partial t}\right].$$

(b) Show that if we define two new potentials

$$\mathbf{A}' = \mathbf{A} + \nabla\chi,$$

$$\Phi' = \Phi - \frac{\partial\chi}{\partial t},$$

where χ is an arbitrary scalar function of position and time, then

$$\mathbf{B} = \nabla \times \mathbf{A}',$$

$$\mathbf{E} = -\nabla\Phi' - \frac{\partial\mathbf{A}'}{\partial t}.$$

That is, the fields $\mathbf{E}$ and $\mathbf{B}$ are not modified by the change in the potentials $\mathbf{A}$ and Φ. The change from $(\mathbf{A}, \Phi)$ to $(\mathbf{A}', \Phi')$ is called a *gauge transformation*.

(c) Show that if we require χ to satisfy the equation

$$\nabla^2\chi - \epsilon_0\mu_0 \frac{\partial^2\chi}{\partial t^2} = -\left[\nabla \cdot \mathbf{A} + \epsilon_0\mu_0 \frac{\partial\Phi}{\partial t}\right],$$

then

$$\nabla \cdot \mathbf{A}' + \epsilon_0\mu_0 \frac{\partial\Phi'}{\partial t} = 0.$$

(d) If χ satisfies the equation given in (c), show that A' and Φ' satisfy the equations

$$\nabla^2\Phi' - \epsilon_0\mu_0 \frac{\partial^2\Phi'}{\partial t^2} = -\rho/\epsilon_0$$

and

$$\nabla^2\mathbf{A}' - \epsilon_0\mu_0 \frac{\partial^2\mathbf{A}'}{\partial t^2} = -\mu_0\mathbf{J}.$$

The point to all of this is that we can make a gauge transformation [as in (b)], impose the condition given in (c), and thereby obtain a scalar and a vector potential which satisfy the equations in part (d), which are wave equations with source terms proportional to ρ and $\mathbf{J}$.

IV–28 The equation of motion of an ideal fluid can be written

$$\iiint_V \rho\mathbf{f}_{ext} \, dV - \iint_S \hat{n}p \, dS = \iiint_V \rho\left[\frac{\partial\mathbf{v}}{\partial t} + (\mathbf{v} \cdot \nabla)\mathbf{v}\right] dV$$

where V is the volume of the fluid and S is its surface. Here $\mathbf{f}_{ext}(x, y, z)$

is the external force per unit mass acting on the fluid, $p(x, y, z)$ is the pressure of the fluid, and $\rho(x, y, z)$ is its density, all at a point (x, y, z) in the fluid, and $\mathbf{v}(x, y, z, t)$ is the velocity of the fluid at the point (x, y, z) and at time t.

(a) Use the form of the divergence theorem given in Problem IV–25(c) to rewrite the equation of motion of an ideal fluid in the form

$$\mathbf{f}_{\text{ext}} - \frac{1}{\rho} \nabla p = \frac{\partial \mathbf{v}}{\partial t} + (\mathbf{v} \cdot \nabla)\mathbf{v}.$$

(b) Show that in the static case ($\mathbf{v} = 0$), the equation of motion becomes

$$\mathbf{f}_{\text{ext}} = (1/\rho)\nabla p.$$

(c) Consider a column of incompressible fluid oriented vertically parallel to the z-axis as shown in the figure. Assuming that the only external force acting on the fluid is the downward uniform gravitational attraction of the earth, apply the equation for the static case given in (b) to show that

$$p = p_0 - \rho g z$$

where g is the acceleration due to gravity and p_0 is a constant.

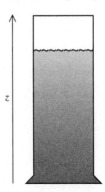

Solutions
to Problems

One must learn by doing the thing; for
though you think you know it, you have no
certainty until you try.

Sophocles

Chapter I

3. (a) $(\mathbf{i}x + \mathbf{j}y)/\sqrt{x^2 + y^2}$.
   (b) $(\mathbf{i} + \mathbf{j})(x + y)^2/\sqrt{2}$.
   (c) $-\mathbf{i}y + \mathbf{j}x$.
   (d) $(\mathbf{i}x + \mathbf{j}y + \mathbf{k}z)/\sqrt{x^2 + y^2 + z^2}$.

4. (a) $(a^2 \cos^2 \omega t + b^2 \sin^2 \omega t)^{1/2}$.
   (b) $-\mathbf{i}\omega a \sin \omega t + \mathbf{j}\omega b \cos \omega t$ (velocity).
       $-\mathbf{i}\omega^2 a \cos \omega t - \mathbf{j}\omega^2 b \sin \omega t$ (acceleration).

5. $-\dfrac{\mathbf{i}}{2\pi\epsilon_0} \dfrac{1}{(y^2 + 1)^{3/2}}$.

6. In the following, c is an arbitrary constant.
   (a) $x^2 - y^2 = c$.   (b) $y = x + c$.

(c) $xy = c$.

(d) $y = c$.

(e) $x = c$.

(f) $x^2 - y^2 = c$.

(g) $y = \frac{1}{2}x^2 + c$.

(h) $y = ce^x$.

Chapter II

1. (a) $(\mathbf{i} + \mathbf{j} + \mathbf{k})/\sqrt{3}$.
   (b) $-(\mathbf{i}x + \mathbf{j}y - \mathbf{k}z)/\sqrt{2}z$.
   (c) $\mathbf{i}x + \mathbf{k}z$.
   (d) $(-2\mathbf{i}x - 2\mathbf{j}y + \mathbf{k})\sqrt{1 + 4z}$.
   (e) $(\mathbf{i}x + \mathbf{j}y + \mathbf{k}a^2z)/a\sqrt{1 + (a^2 - 1)z^2}$.

3. $[-\mathbf{i}(\partial g/\partial x) - \mathbf{k}(\partial g/\partial z) + \mathbf{j}]/\sqrt{1 + (\partial g/\partial x)^2 + (\partial g/\partial z)^2}$
   $$\text{for } y = g(x, z).$$

   $[-\mathbf{j}(\partial h/\partial y) - \mathbf{k}(\partial h/\partial z) + \mathbf{i}]/\sqrt{1 + (\partial h/\partial y)^2 + (\partial h/\partial z)^2}$
   $$\text{for } x = h(y, z).$$

4. (a) $\sqrt{3}/6$.

   (b) $\dfrac{\pi}{2}(\sqrt{5} - 1)$.

   (c) $\pi/2$.

5. (a) 0.

   (b) $2\pi a^3$.

   (c) $3\pi/2$.

6. $4\pi R^2 \sigma_0/3$.

7. $16\pi R^4 \sigma_0/15$

8. 0.

9. $\pi r^2 \lambda h \epsilon_0$.

10. (a) 0.

    (b) $4\pi R^2 h \ln R$.

    (c) $4\pi R^3 e^{-R^2}$.

    (d) $[E(b) - E(0)]b^2$.

11. (a) $\mathbf{E} = \sigma \mathbf{i}/2\epsilon_0$, $x > 0$, and $-\sigma \mathbf{i}/2\epsilon_0$, $x < 0$.
    (b) $\mathbf{E} = \rho_0 b \mathbf{i}/\epsilon_0$, $x > b$; $\rho_0 x \mathbf{i}/\epsilon_0$, $-b \le x \le b$;
        and $-\rho_0 b \mathbf{i}/\epsilon_0$, $x < -b$.
    (c) $\mathbf{E} = \pm (\rho_0 b/\epsilon_0)(1 - e^{-|x|/b})\mathbf{i}$ (+ for $x > 0$, $-$ for $x < 0$)

12. (a) $\mathbf{E} = (\lambda/2\pi\epsilon_0)\hat{\mathbf{e}}_r/r$
    (b) $\mathbf{E} = (\rho_0 b^2/2\epsilon_0)\,\hat{\mathbf{e}}_r/r$, $r \ge b$, and $(\rho_0 r/2\epsilon_0)\,\hat{\mathbf{e}}_r$, $r \le b$.
    (c) $\mathbf{E} = (\rho_0 b^2/\epsilon_0)(1/r)[1 - (1 + r/b)e^{-r/b}]\hat{\mathbf{e}}_r$.

13. (a) $\mathbf{E} = \begin{cases} (b^3 \rho_0/3\epsilon_0)\hat{\mathbf{e}}_r/r^2, & r > b, \\ (\rho_0/3\epsilon_0)r\hat{\mathbf{e}}_r, & r \le b. \end{cases}$

    (b) $\mathbf{E} = (b^3 \rho_0/\epsilon_0)(1/r^2)[2 - (r^2/b^2 + 2r/b + 2)e^{-r/b}]\hat{\mathbf{e}}_r$.

    (c) $\mathbf{E} = \begin{cases} (\rho_0/3\epsilon_0)r\hat{\mathbf{e}}_r, & r < b, \\ (1/3\epsilon_0)(1/r^2)[b^3\rho_0 + (r^3 - b^3)\rho_1]\hat{\mathbf{e}}_r, & b \le r \le 2b, \\ (b^3/3\epsilon_0)(1/r^2)(\rho_0 + 7\rho_1)\hat{\mathbf{e}}_r, & r > 2b. \end{cases}$

    The field is zero for $r > 2b$ if $\rho_1 = -\rho_0/7$. The total charge is then zero.

14. (a) $2(x + y + z)$.

    (b) 0.

    (c) $-(e^{-x} + e^{-y} + e^{-z})$.

    (d) $2z$.

    (e) $-y/(x^2 + y^2)$.

    (f) 0.

    (g) 3.

    (h) 0.

15. Surface integral equals $2s^3(x_0 + y_0 + z_0)$ for the function of Problem II–14a.

    Surface integral equals 0 for function of Problem II–14b.

    Surface integral equals $s^2(e^{-s/2} - e^{s/2})(e^{-x_0} + e^{-y_0} + e^{-z_0})$ for function of (II–14C).

16. (b) $\nabla \cdot \mathbf{G} = 0$.

22. $f(r) = \text{constant}/r^2$.

23. (a) $3b^3$.

    (b) $3\pi R^2 h/4$.

    (c) $4\pi R^4$.

24. (b) $\pi R^2 B$.

Chapter III

3. (a) $2(-\mathbf{i}y + \mathbf{j}z + \mathbf{k}x)$.

    (b) $5\mathbf{j}x$.

    (c) $\mathbf{i}e^{-z} + \mathbf{j}e^{-x} + \mathbf{k}e^{-y}$.

    (d) 0.

    (e) $-\mathbf{i}x - \mathbf{j}y + 2\mathbf{k}z$.

    (f) $2(\mathbf{i}y - \mathbf{j}x)$.

    (g) $\mathbf{i}z - \mathbf{k}x$.

    (h) 0.

4. Line integral equals $2x_0 s^2$ for function of Problem III–3a.

    Line integral equals 0 for function of Problem III–3b.

    Line integral equals $s(e^{s/2} - e^{-s/2})e^{-y_0}$ for function of Problem III–3c.

    Line integral equals 0 for function of Problem III–3d.

5. (a) $a^2/2 + a^3/3$

    (b) $\dfrac{2}{\sqrt{3}} \left(\dfrac{1}{2} + \dfrac{a}{3} \right) \rightarrow \dfrac{1}{\sqrt{3}}$ as $a \rightarrow 0$

    (c) $\hat{n} \cdot \nabla \times \mathbf{F} = \dfrac{1}{\sqrt{3}} (1 + 2y) = \dfrac{1}{\sqrt{3}}$ at $y = 0$

13. (d) and (h).

15. (a) Line integral and surface integral equal 1.

    (b) Line integral and surface integral equal $-3\pi/4$.

    (c) Line integral and surface integral equal $-2\pi R^2$.

19. $3/e$.

25. If $\nabla \cdot \mathbf{G} = 0$, then $\mathbf{G} = \nabla \times \mathbf{H}$.

    (a) $\mathbf{H} = \frac{1}{2}\mathbf{j}x^2 + \mathbf{k}[\frac{1}{2}y^2 - (x - x_0)z]$.

    (b) $\mathbf{H} = \mathbf{j}B_0 x$.

    (c) $\nabla \cdot \mathbf{G} \neq 0$.

    (d) $\mathbf{H} = -\mathbf{j}(x - x_0)z + \mathbf{k}(x + x_0)y$.

    (e) $\nabla \cdot \mathbf{G} \neq 0$.

    [*Note:* Your results may differ from these by additive constants.]

28. (a) $\oint_C \mathbf{H} \cdot \hat{\mathbf{t}} \, ds = \iint_S \mathbf{G} \cdot \hat{\mathbf{n}} \, dS$ where S is a capping surface of the closed curve C.

29. (d) Surface and volume integral each equal $\mathbf{i} - \mathbf{j} - \mathbf{k}$.

Chapter IV

1. (a) (i) $\mathbf{F} = \mathbf{i}yz + \mathbf{j}xz + \mathbf{k}xy$.

   (ii) $\mathbf{F} = 2(\mathbf{i}x + \mathbf{j}y + \mathbf{k}z)$.

   (iii) $\mathbf{F} = \mathbf{i}(y + z) + \mathbf{j}(x + z) + \mathbf{k}(x + y)$.

   (iv) $\mathbf{F} = 6\mathbf{i}x - 8\mathbf{k}z$.

   (v) $\mathbf{F} = -\mathbf{i}e^{-x} \sin y + \mathbf{j}e^{-x} \cos y$.

4. (a) (i) Not path-independent.

   (ii) $\psi = c_z = $ const.

   (iii) $\psi = xyz + $ const.

   (iv) $\psi = \frac{1}{2}(x^2 + y^2 + z^2) + $ const.

   (v) Not path-independent.

   (b) (i) $\psi = \ln r + $ const.

   (ii) $\psi = \frac{2}{3}r^{3/2} + $ const.

13. (a) $\rho = 3g\epsilon_0$.

    (b) $\Phi = -\frac{1}{2}g(x^2 + y^2 + z^2)$.

16. (a) Add V_0 to the result obtained in the text.

17. (a) $\Phi(r, \theta) = -E_0 r(1 - R^3/r^3) \cos \theta$ where the sphere is centered at the origin and the two plates are parallel to the xy-plane and situated at $z = \pm s/2$.

    (c) Add V_0 to the result given in part (a).

19. (a) Move in the direction $\pm(\mathbf{i}b - \mathbf{j}a)/\sqrt{a^2 + b^2}$.

    (b) Move in the direction of the gradient:
    $-(\mathbf{i}a + \mathbf{j}b)/\sqrt{r^2 - a^2 - b^2}$.

    (c) Move in the direction opposite to the gradient:
    $(\mathbf{i}a + \mathbf{j}b)/\sqrt{r^2 - a^2 - b^2}$.

25. (e) Surface integral and volume integral each equal $\pi\mathbf{k}$.

Bibliography

Arfken, G. *Mathematical Methods for Physicists.* New York: Academic Press, Inc., 1968.

Aris, R. *Vectors, Tensors,* and the Basic Equations of Fluid Mechanics. Englewood Cliffs, N.J.: Prentice-Hall, Inc., 1962.

Barnett, R. A., and Fujii, J. N. *Vectors.* New York: John Wiley & Sons, Inc., 1963.

Buck, R. C. *Advanced Calculus.* New York: McGraw-Hill Book Company, Inc., 1965.

Carslaw, H. S., and Jaeger, J. C. *Conduction of Heat in Solids.* New York: Oxford University Press, 1947.

Churchill, R. V. *Fourier Series and Boundary Value Problems.* New York: McGraw-Hill Book Company, Inc., 1963.

Crank, J. *The Mathematics of Diffusion.* New York: Oxford University Press, 1956.

Dennemeyer, R. *Introduction to Partial Differential Equations and Boundary Value Problems.* New York: McGraw-Hill Book Company, Inc., 1968.

Dixon, C. *Applied Mathematics of Science and Engineering.* New York: John Wiley & Sons, Inc., 1971.

Green, B. A., Jr. *Vector Calculus.* New York: Appleton-Century-Crofts, 1967.

Hildebrand, F. B. *Advanced Calculus for Applications.* Englewood Cliffs, N.J.: Prentice-Hall, Inc., 1962.

Jackson, J. D. *Classical Electrodynamics.* New York: John Wiley & Sons, Inc., 1962.

Kaplan, W. *Advanced Calculus.* Boston: Addison-Wesley Publishing Company, Inc., 1952.

Karamcheti, K. *Vector Analysis and Cartesian Tensors.* San Francisco: Holden-Day, 1967.

Kellogg, O. D. *Foundations of Potential Theory.* Berlin: Springer-Verlag, 1929.

Bibliography

Kip, A. F. *Fundamentals of Electricity and Magnetism.* New York: McGraw-Hill Book Company, Inc., 1969.

Kline, M. *Calculus: An Intuitive and Physical Approach.* New York: John Wiley & Sons, Inc., 1967.

Landau, L. D., and Lifshitz, E. M. *Fluid Mechanics.* New York: Pergamon Press, Inc., 1959.

McLeod, E. B., Jr. *Introduction to Fluid Dynamics.* New York: Pergamon Press, Inc., 1963.

Pipes, L. A., and Harvill, L. R. *Applied Mathematics for Engineers and Physicists.* New York: McGraw-Hill Book Company, Inc., 1970.

Purcell, E. M. *Electricity and Magnetism.* New York: McGraw-Hill Book Company, Inc., 1965.

Sokolnikoff, I. S., and Redheffer, R. M. *Mathematics of Physics and Modern Engineering.* New York: McGraw-Hill Book Company, Inc., 1966.

Thomas, G. B., Jr. *Calculus and Analytic Geometry.* Reading, Mass.: Addison-Wesley Publishing Company, Inc., 1968.

Index

164

	CARTESIAN	CYLINDRICAL

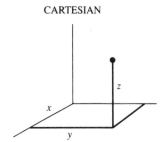

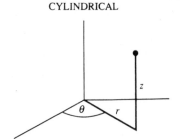

DIVERGENCE
div **F**
$\nabla \cdot \mathbf{F}$

$$\frac{\partial F_x}{\partial x} + \frac{\partial F_y}{\partial y} + \frac{\partial F_z}{\partial z}$$

$$\frac{1}{r}\frac{\partial}{\partial r}(rF_r) + \frac{1}{r}\frac{\partial F_\theta}{\partial \theta} + \frac{\partial F_z}{\partial z}$$

GRADIENT
grad f
∇f

$$(\nabla f)_x = \frac{\partial f}{\partial x}$$

$$(\nabla f)_r = \frac{\partial f}{\partial r}$$

$$(\nabla f)_y = \frac{\partial f}{\partial y}$$

$$(\nabla f)_\theta = \frac{1}{r}\frac{\partial f}{\partial \theta}$$

$$(\nabla f)_z = \frac{\partial f}{\partial z}$$

$$(\nabla f)_z = \frac{\partial f}{\partial z}$$

CURL
curl **F**
$\nabla \times \mathbf{F}$

$$(\nabla \times \mathbf{F})_x = \frac{\partial F_z}{\partial y} - \frac{\partial F_y}{\partial z}$$

$$(\nabla \times \mathbf{F})_r = \frac{1}{r}\frac{\partial F_z}{\partial \theta} - \frac{\partial F_\theta}{\partial z}$$

$$(\nabla \times \mathbf{F})_y = \frac{\partial F_x}{\partial z} - \frac{\partial F_z}{\partial x}$$

$$(\nabla \times \mathbf{F})_\theta = \frac{\partial F_r}{\partial z} - \frac{\partial F_z}{\partial r}$$

$$(\nabla \times \mathbf{F})_z = \frac{\partial F_y}{\partial x} - \frac{\partial F_x}{\partial y}$$

$$(\nabla \times \mathbf{F})_z = \frac{1}{r}\frac{\partial}{\partial r}(rF_\theta) - \frac{1}{r}\frac{\partial F_r}{\partial \theta}$$

LAPLACIAN
$\nabla^2 f$

$$\frac{\partial^2 f}{\partial x^2} + \frac{\partial^2 f}{\partial y^2} + \frac{\partial^2 f}{\partial z^2}$$

$$\frac{1}{r}\frac{\partial}{\partial r}\left(r\frac{\partial f}{\partial r}\right) + \frac{1}{r^2}\frac{\partial^2 f}{\partial \theta^2} + \frac{\partial^2 f}{\partial z^2}$$